# THE ROLE OF THE STATE AND INDIVIDUAL IN SUSTAINABLE LAND MANAGEMENT

# The Role of the State and Individual in Sustainable Land Management

*Edited by*

ROBERT W. DIXON-GOUGH
*University of East London, UK*

PETER C. BLOCH
*University of Wisconsin-Madison, USA*

Routledge
Taylor & Francis Group
LONDON AND NEW YORK

First published 2006 by Ashgate Publishing

Reissued 2018 by Routledge
2 Park Square, Milton Park, Abingdon, Oxon, OX14 4RN
711 Third Avenue, New York, NY 10017, USA

*Routledge is an imprint of the Taylor & Francis Group, an informa business*

First issued in paperback 2018



A Library of Congress record exists under LC control number: 2005034783



ISBN 13: 978-0-815-39816-5 (hbk)
ISBN 13: 978-1-138-62275-3 (pbk)
ISBN 13: 978-1-351-14548-0 (ebk)

# Contents

# List of Figures and Tables

# Chapter 1

# Introduction

Robert Dixon-Gough

The role of the State and the individual, particularly in aspects relating to the land, are inexorably intertwined. The State provides the legislation relating to the use and even ownership of the land and controls that legislation. However, it is frequently the individual who physically shapes the characteristics of the landscape, often with the encouragement of the State. Historically, the nature of the 'State' and the territories it controls has varied from local, through regional, to national and now international. The method of control has also varied through history through power, empowerment, dictatorship, to democracy. All stages in this evolution have left patterns that have shaped the landscape, and formulated the methods and practices of land administration and land management. One of the most important facets of land administration and land management is land ownership.

It is virtually impossible to determine with any degree of certainty the historical extent of land management and land ownership since many of the oldest forms of evidence have been obliterated by subsequent developments or are insufficiently understood. It is, however, acknowledged by archaeologists that some of the best preserved and most extensive evidence of settled landscapes may be found in the British Isles, particularly in the chalk uplands of central, southern England (Bradley, 1994). These patterns date back to the Neolithic and earlier Bronze Age periods and might have been ceremonial centres of earlier periods that had been incorporated into new land divisions (Bradley, 1981). Whilst some of the more extensive linear boundaries might have been symbols of empowerment by the 'State', others form enclosure that is related more to ownership or stewardship than empowerment. Ironically, the evidence of these patterns was far more extensive when the idea of National Parks in England and Wales was first discussed. Archaeologists recommended that the remains of earthworks on the Marlborough Downs (central southern England) were so well preserved (in contrast to those on Salisbury Plain, which was then and still is a military training area) that the area should be protected (Evans, 1994).

This is no longer true. The Marlborough Downs never became a National Park and the archaeological remains have largely been destroyed by agriculture, sustained through subsidies by the State, whilst on Salisbury Plain (also subsidised by the State) much of the archaeology is intact and has not been touched by modern farming techniques.

One of the common features relating to evidence of settlements and society is that they lie in upland areas. Whilst the earliest evidence of the impact of the state and the individual appears in central southern England, it is the uplands and marginal regions of the British Isles that provide evidence of the gradual evolution of land management practices. For example, Rippon (2000a; 2000b) provides evidence of a more formalised interrelationship between the State and land ownership in his discussion of the landscapes of Romano Britain and its transition though the early medieval period whilst Bond (2000) considers the implications of large, non-state land owners, in the form of monasteries and other ecclesiastical bodies on the shaping of the landscape. Whyte et al. (2004) provide a detailed overview of the evolution of society, the landscape, and the environment in upland Britain, which again emphasises the message that the greatest evidence of change in the role of the State and the individual upon land management, in general, may be found away from the more populated and intensively used regions. This is, in effect, a truism but nevertheless the evidence that is available to mark those changes is impressive. For examples, see Dodgshon (2004); Straughton (2004) and Whyte (2004).

This concept of change and the evolution of the role of the State and the individual in land management is extended by Arnison who considers the role of property rights as a social institution. Whilst property ownership is an accepted institution in most western States, there are many regions of the world, in particular some countries of the former Soviet Union in which private, as distinct from State, ownership is still a novelty.

Whilst accepting the concept of property ownership, Dixon-Gough extends this to responsibility through the case study of Grange-over-Sands and the water catchment area of the River Kent. Here, he demonstrates that through a combination of private ownership initiatives and state-supported land and agrarian reform programmes, the landscape in transformed in a non-sustainable manner that has direct implications upon adjacent landscape features, in this case, the foreshore of Grange-over-Sands, which has become covered by saltmarsh.

Land management is also the theme of the chapter by Morad who examines a similar concept to that of Dixon-Gough, namely the interrelationship between land management and ownership but extends this to biodiversity in New Zealand. Land ownership in New Zealand have been under scrutiny as a result of a rapid loss of native biodiversity and urban expansion, coupled to a rise and acceptance of Maori land claims. These claims, recognised as being valid and justifiable, have been the subject of Resource Land Management Acts that have appeared to reconcile the incompatible goals between the state and the individual by being legally sound, culturally appropriate and environmentally sustainable. These serve as an exemplar to other states throughout the world.

In a similar context to Morad, Evans investigates the complex processes associated with deforestation in the neotropics and discusses the challenges of developing a programme of sustainable land management based upon both the intervention of the state and through the responsibility of the individual. In this chapter, Evans stresses the need to balance the multiple demands of society and natural resources, with those of conservation and production to develop policies based upon sustainable outcomes.

One of the roles of the State is the protection of individual property rights and ownership. This can become very contentious in marginal areas, where there may be an impact upon the rights and property of the individual, and environmental phenomena that, in the opinion of that owner, can be controlled. An example of this is the coastal zone, particularly the soft coastlines that are under threat through changing environmental and climatic conditions. Wynn describes a situation in the Blackwater Estuary in which the situation is the direct opposite to that described earlier by Dixon-Gough. In the case of Grange-over-Sands, there is net accretion resulting in the formation of saltmarshes, whereas in the Blackwater Estuary, the problem is that of erosion and coastal defences from the threat of coastal flooding. Wynn describes the balance that must be made between protecting the environment and habitats, and protecting individual properties, a situation where the state has often to intervene to protect the environment rather than the rights and property of individuals. In so doing, Wynn recognises the need for an internationally accepted role in coastal management, directed on a regional basis rather than by local government or the state.

In contrast to the specific roles of the state and the individual considered in by Dixon-Gough, Morad, and Wynn, Bledsoe considers the administrative institutions that serve general land-related functions. These institutions provide a wide range of functions include those normally controlled by the state such as land titling, planning, surveying and mapping, through to the more commercial activities like credit and mortgages, insurance, and the sale of real property. Whilst there may be a clear division between the role of the state and 'individual' in many regions of the world, the differences are becoming blurred as many government institutions are becoming more cost-accountable and are being organised along commercial lines. Bledsoe considers the characteristics of land-related administrations with respect to productive land development in Central and Eastern Europe and concludes that consensus building can be as valuable as capital investments in technology or facilities.

The need to develop consensus, particularly between the State and the individual is developed by Dixon-Gough and Molobeng through an examination of the procedures adopted in Botswana in the managing and upgrading of informal settlements. This is particularly pertinent since most countries around the world have experienced the effects of mass migrations from rural to urban areas, and those who have fled conflict zones. Unfortunately, they are a common feature of developing countries and the treatments of those settlements range from periodic destruction by the 'state', to help in the provision of essential services. This chapter examines the policies adopted in Francistown, Botswana, in the formalisation of those settlements, together with granting the occupants with a form of land title that provides the rights and responsibilities of individual ownership upon the family unit.

Individual ownership and property rights is also the subject of the chapter by Bloch et al in the evaluation of the role of USAID in contributing towards the development of those rights. In particular, this chapter examines the developing role of land markets and the economic growth. It hypothesises that providing property owners have the incentive to invest in their properties, they will increase their investment and may wish to acquire further land, through entrepreneurs, to increase their ownership, which in turn enhances their ability to borrow more and increase the levels of improvement. However, the key to the sustainability of this hypothesis is stability and the guarantee by the state of individual property rights.

Whilst property rights within a country may be supported and guaranteed by the state, and procedures developed to guarantee the efficient transactions of those properties, real property should be easily transferable to make the real estate market more efficient. In some countries, the procedures are simple whilst in others the process is more problematical. This is largely due to way in which their legal frameworks and institutions have evolved. Vaskovitch et al compare the purchase and subdivision processes in three jurisdictions, those of Denmark, England and Wales, and Belarus. This is part of a much larger study, which has been undertaken by the COST Action G9, 'Modelling Real Property Transactions' in an attempt to increase the transparency of those processes across Europe.

As in the case of Belarus, the real estate markets of Kyrgystan have had to be developed since 1991. Since the real estate markets in Kyrgystan had not existed prior to 1991, there was a lack of real estate specialists and furthermore, much of the land reform was based upon programmes of agrarian reform rather than real estate. Bloch describes the evolution of these processes and concludes that whilst land reform is almost complete, and the rights of the owners protected by the state, regulation of the property markets may be needed to ensure long-term equity and sustainability.

In the penultimate chapter, Dixon-Gough reviews one of the problems of land ownership, that of land fragmentation through either inheritance or land redistribution. The former is a factor of continued individual land ownership whilst the latter is largely the result of governmental policies. The problem of land fragmentation is most serious in rural area where property is sub-divided as a result of inheritance. Over a period of two or three generations, the plot size may be too small to be economically farmed. Furthermore, the very process of land redistribution may result in the proliferation of small plots, unsuited to the demands of modern agriculture. The normal tool used to overcome these problems is land consolidation or land readjustment, and the role of these tools – administered by the state – upon the individual land owners is discussed and evaluated.

Whilst land consolidation is a tool used to consolidate fragmented holdings, the land tenure reforms introduced by the devolved Scottish Parliament might conceivably lead to a greater level of fragmentation of holdings in parts of Scotland. Andrew discusses the rapid evolution of reforms to the property laws of Scotland, which came into force at the end of November 2004. Although these reforms addressed virtually all issues of property ownership in Scotland, one of the most emotive has been the empowerment of rural communities to purchase their land, and the (sometimes valuable) interests of that land, even when the landlord may be unwilling so to do. Whilst this will lead to fragmentation, it will also lead to a greater degree of social and economic justice, particularly within the highlands and islands of Scotland.

## References


Bond, J. 2000. Landscapes of monasticism. In: Hooke, D. (ed.) *Landscape: the Richest Historical Record*, 63-74, The Society for Landscape Studies Supplementary Series 1.

Bradley, R., 1981. The evolution of later Bronze Age settlements. In: Barrett, J. and Bradley, R. (eds.) *Settlement and Society in the British later Bronze Age*, BAR, 83, 77-100, Oxford.

Bradley, R., 1994. The background to the enquiry. In: Bradley, R., Entwistle, R. and Raymond, F. (eds.) *Prehistoric Land Divisions on Salisbury Plain. The Work of the Wessex Linear Ditches Project*, 3-16, English Heritage, London.

Dodgshon, R.A., 2004. The Scottish Highland before and after the Clearances: an ecological perspective. In: Whyte, I.D. and Winchester, A.J.L. (eds.) *Society, Landscape and Environment in Upland Britain*, 67-78, The Society for Landscape Studies Supplementary Series 2.

Evans, D.M., 1994. Preface. In: Bradley, R., Entwistle, R. and Raymond, F. (eds.) *Prehistoric Land Divisions on Salisbury Plain. The Work of the Wessex Linear Ditches Project*, 1-2, English Heritage, London.

Rippon, 2000a. The Romano-British exploitation of coastal wetlands: survey and excavation on the North Somerset Levels, 1993-97, *Britannia*, **31**, 69-200.

Rippon, S. 2000b. Landscapes in transition: the later Roman and early medieval periods. In: Hooke, D. (ed.) *Landscape: the Richest Historical Record*, 47-62, The Society for Landscape Studies Supplementary Series 1.

Staughton, E.A., 2004. Beyond enclosure: upland common land in England and Wales since 1800. Whyte, I.D. and Winchester, A.J.L. (eds.) *Society, Landscape and Environment in Upland Britain*, 89-98, The Society for Landscape Studies Supplementary Series 2.

Whyte, I.D., 2004. The landscape and environmental impact of mining and quarrying in upland Britain. Whyte, I.D. and Winchester, A.J.L. (eds.) *Society, Landscape and Environment in Upland Britain*, 111-124, The Society for Landscape Studies Supplementary Series 2.

Whyte, I.D. and Winchester, A.J.L. (eds.) *Society, Landscape and Environment in Upland Britain*, 67-78, The Society for Landscape Studies Supplementary Series 2.

Chapter 2

# Land Ownership and Personal Empowerment: A Review of Some Aspects of Current Land Reforms in the Light of the Lessons of Modern Social History

Chris Arnison

## Introduction

The role of property rights in land as a social institution encountered in most cultures and social systems in the modern era, and in many in the pre-modern era, has been well documented (e.g. by Beaglehole, 1931; Davis, 1948; Ardrey, 1967; Reeve, 1986). The importance of private territorial rights, and the concomitant deferential social postures of lesser 'territors' or 'non-territors',[1] has been noted in contexts as diverse as the relationship between municipal housing authorities and their tenants (Reynolds, 1987), the voting behaviour of farm workers in East Anglia (Newby, 1977), the reactions of householders to those who have burgled their houses (Trasler, 1982), and the difference in performance of 'home' and 'away' sports teams (Irving and Goldstein, 1990).

There seems to be general agreement on at least two points among all writers on the broad topic of private property rights, embracing both real and personal property. First, that they are a universal feature of all human societies, both ancient and modern – that they are a true social institution – and second that they are beneficial both to the psychological health of the individual and to the social coherence and stability of the human group within which the individual lives. Laurence Hobhouse, one of the first professors of sociology in the UK, put it this way:

> Some measure of property appears ... to be the essential basis of liberty; and the sense of freedom in enjoyment ranks along with the sense of security and

[1] The terms were coined by the present writer (Arnison, 1989) because current English has no simple expression for 'a person or body endowed with recognised territorial rights', other than the word 'proprietor', which carries too much of a European legal connotation.

> permanence among the complex constituents of the pride and joy of ownership … Material things that a man can count on as his own, that he can leave and return to, that he can use at his will, are (we admit) the basis of a purposeful life and therefore of a rational and harmonious development of personality (Hobhouse, 1913).

The rational argument for private property in land and buildings to be restored or distributed to citizens of the Former Soviet Union (FSU) and the new Eastern European democracies is based chiefly on economic grounds: land and buildings are an important form of wealth, which can stimulate market activity, including investment and development. Therefore a way must be found for such rights to become tradeable, and that can only be done if they are owned by someone, whether a natural or a legal person.

There is, however, also a case for private ownership based not on rational grounds but on socio-biological grounds – drawing on research into human behaviour. Strong claims have been made for human territoriality to be regarded as a true instinct (e.g. by Ardrey op. cit.; Malmberg, 1980) and therefore to be outside the scope of rational conduct. Such claims are given weight by the universal experience of neighbour disputes that appear to have an importance to the protagonists out of all proportion to their physical scale in terms of actual space or to the economic value of that space. Indeed, press reports of neighbours who beggar both themselves and each other through protracted litigation appear regularly wherever private property exists.

Persuasive explanations, both theoretical and empirical, for man's appetite for and attitude towards private property are provided by the science of social anthropology: man is a social animal, not a solitary one, and is in nature a member of a clan group, more analogous to a pack of wolves or a pride of lions than to a herd of deer (Beaglehole, 1931; Davis, 1948). Such clans function most effectively when there is a clear clan leader whose status, in man at least, requires externally visible proofs. One of the most potent of these is a conspicuous personal private space from which the leader can exclude all others. In modern life, this is invariably a home with a visible boundary. The bigger the space around the home, as well as the size of the home itself, the greater the quantum of personal power, and the greater the deference afforded to the holder of such a power-symbol. In many cases, a house becomes a virtual castle, both in terms of its solidity and defensive equipment (high boundary walls, intruder alarms, remotely controlled gates) and its physical size and architectural features. Individuals who want status in their clan and the deference that goes with it are therefore always alert for opportunities for acquiring a territory, which will confer such status. The universal application of this rather simple model of modern human behaviour is attested by, for example, the rush to purchase confiscated ecclesiastical estates by the wealthy lawyers and merchants of Henry VIII's England as well as by the recent proliferation of ostentatious new private houses in the suburbs of Brunei Darusalaam, funded by the largesse of the present Sultan to those who run his government, or the similar construction of new private houses, with conspicuous external embellishments, by those to whom household plots have been granted, or restored, in the whole sweep of Eastern Europe, from Estonia to Bulgaria.

Ardrey (1970) presents a simple summary of conclusions he has drawn from his own observations and the work of ethologists and social anthropologists:

> The will to power is satisfied by real estate, (which confers) dominance over men ... Territory lingers, but as a symbol of status and, while presenting a degree of territorial security, commands through size and location its far more pressing symbolic contribution to rank.

It is no mere quirk of linguistic history that the landed estate of an English country squire was very often described, in legal documents, family histories or handbooks on estate management (e.g. Clay, 1624), as an 'honourable estate' – meaning an estate which conferred honour (i.e. rank) on its owner. It was (and continues to be today) the ownership of a significant territory rather than a noble title or mere money wealth, which conferred most status and gained most respect from other members of the same 'clan' community. The man who gains status in his social milieu by acquiring the recognised territorial trappings of a higher rank not only gains personal gratification but also respect, deference and even obedience from those who are now below him in rank.

It is instructive to see how Clay introduces his brief manual of estate management, nearly 400 years ago:

> An Honorable estate is reall possessions, which have rights of suits, services, privileges, and Jurisdictions appropriate and belonging to them: that is, for further explanation where any Lord or Owner is....either by inheritance, gift, *purchase, or otherwise*, endowed with Honours, Mannors, lordships, Houses, Lands, Tenements, and such like possessions... For although that honorable titles (in respect of mens persons) doe properly belong unto none but such who by descent, office, place, the Princes free gift, or such like, may justly challenge (i.e.claim) the same; *yet even men of meaner ranke may be seized of such honorable estates.* (italics supplied).[2]

It is the deference and obedience accorded to territors, especially holders of conspicuous territories such as large houses, extensive lands or historic buildings, which is a potential source of concern because of the opportunities such deference gives for manipulation of local administrative or political processes.

This chapter will focus on a number of cases where private property rights are something of a novelty, though still regarded as important, and where a new set of 'haves' is established amongst a larger set (as it ever was!) of 'have nots'. The most conspicuous examples of such cases have arisen in the past ten years in the FSU and its satellite states in Eastern Europe. Some interesting and contrasting examples, however, are found in parts of Africa, for quite different reasons.

Many commentators have described the steps taken by the newly independent states of the FSU, and the newly democratic sovereign governments of satellites such as Bulgaria and Poland, to create, each in its own way, new

---

[2] Clay, op. cit., p. 58.

property-owning democracies (e.g. Thiesenhusen, 1994; Csaki, 1994). In summary those steps, as may be observed them in a dozen or so such countries, comprise:

- creating a constitutional right for citizens to own land and the buildings on it;
- enacting legislation which defines the nature and extent of the rights which citizens can hold over different types of land and/or buildings;
- establishing administrative procedures for citizens to register such rights in a formal Cadastral Register which, having accepted such registration and issued a Certificate of Registration, guarantees the validity of the rights permanently; and
- either confirming that certain existing occupiers of land and/or buildings should be treated as owners or allocating parcels of state or municipal land to deserving persons who also will henceforth be treated as owners.

The sequence given is, in the view of the author, the only logical sequence: but in a number of cases stage four, followed stage one,[3] with the predictable result that a period of uncertainty and confusion has followed before the legal structures which should have been stage two were put in place (Home, 1996). In some cases stage three has been delayed, or bungled, with the result that what should be a tradeable asset remains unsaleable as buyers will not take the risk of buying land with an unregistered title at its full market price and sellers are unwilling to sell below market price. The consequential distortion of normal and orderly market transactions in land rights is, in many cases, severe and seriously hindering the transition from a command economy to a market economy.

The steps set out above constitute the basis for a fundamental re-ordering of social and economic relations within each state. They will transform each economy and stimulate each society in profound and enduring ways – that is what they are intended to do: that is what the people of each state wished to do in exercising their democratic will to change both their governments and their political and economic systems back in the early 1990s. However, if the process is badly conducted, the outcome may be very different from what the people wanted, or from what the new governments intended. Instead of creating new assets, of stimulating initiative and enterprise, of attracting inward foreign investment and of creating a new and efficient source of taxation, the result may be administrative confusion, or even conflict, and electoral frustration and cynicism. If the anticipated benefits of ownership do not become real to the new owners their ability to act as agents in the process of economic reform is negated and their enthusiasm and energy rapidly drains away. Similar distortion of normal market relations in land transaction is evident in many parts of Southern Africa.

There is, however, likely to be another, and in my view more important, outcome if a large new body of property owners is created in a disorderly and

---

3 For example, Lithuania, Georgia and Bulgaria all implemented land allocation or restitution programmes in the years from 1992 to 1995 before enacting legislation which properly defined property rights.

haphazard way. If land is grabbed, so to speak, by those who happen to be in a position to be the first to claim ownership rights in, or even simply take possession of, land parcels, whether in the form of house plots or factory sites or farm land, the new 'owners' will, even before their ownership is actually ratified by the relevant authorities, be able to claim, *and will be given*, 'territor' status and thus assume a role of influence in the community. However illegitimate (or at least precipitate) their claim to land rights, the simple fact that they are visibly exercising proprietorial control over territory will strengthen their social status and weaken the resolve of anyone who may wish to overturn their claim. As Altman found in his work on territorial behaviour in the 1970s, there is a deeply ingrained human tendency to defer to the occupier of a private personal territory, whether the occupation is as a home, or a farm or for business purposes. The aphorism 'possession is nine tenths of the law' is in reality no more than a statement of normal human behaviour: it is *normal* to assume that there is a 90 per cent chance that occupation means rightful ownership and therefore to be hesitant about challenging such occupation.

There is also an important significance to the word 'rightful' in that last sentence: it confers respectability on an occupier per se, irrespective of his or her real qualities or defects. And further, in the context of rapidly changing local communities in the FSU and Eastern Europe, there is evidence already that some of the more quick-footed of the former leaders as well as some of the less scrupulous new entrepreneurs have contrived to be among the new property owners and therefore to hold new positions in a local community whose members are going to be, naturally, predisposed to defer to them, in local economic decisions, administrative actions and political debates, in just the same way as all European and many Asian communities have deferred to their landowner members for many centuries.

However, the worrying difference is that many of the *historic* landowners to whom deference in local community matters was naturally shown behaved in a notably responsible way in exercising the authority which such deference gave to them; they saw themselves as stewards as well as owners; as trustees with responsibilities to future generations rather than as proprietors entitled to exploit resources to the limit. Certainly, their gaze was focused on future generations of their own family or clan, but so what, if the result is care for both the land and the community, which it serves. In contrast, many of the new owners appear to exhibit, as owners, precisely the qualities that enabled them to become owners quickly – selfishness and disregard for others.

This brings us to Southern Africa, where the current concern for land reform, ongoing now for twenty years or so, also contains a clear conflict between exploitation and stewardship. Here the polarities are between communal ownership by a village – in effect by a clan of a tribe – and freehold ownership by individuals with freedom to use, abuse, and dispose of land as they wish. Between those two poles is a range of intermediate arrangements – short leases, long leases, certificates of rights (in effect no more than a licence to occupy) or customary land grants – all of which fail in one way or another to satisfy either the aspirations of individuals or the cultural traditions of the tribe or the policies of the government.

In a very recent paper, Faustin Kalabamu has concisely summarised a complex debate in Botswana in which the cultural incompatibilities of European Roman law as imported or adopted by the Dutch and English colonisers on the one hand and traditional African communal tenure on the other are being painstakingly smoothed out, not yet successfully, but at least in a calm and measured manner (Kalabamu, 2000). Here, the essence of the conflict is the contrast between the concept of the tribe or clan as an entity in itself, with its own traditional territory, administered by the clan leaders, the Headman and the Village Council of Elders, acting together as custodian of the communal heritage and doing the best they can to accommodate new families and changing demographic structures,[4] and the same 'modernising' drive to give land rights to individuals in the expectation that enterprise and initiative will be stimulated as has motivated the land reforms in the FSU and Eastern Europe. Here too local politicians, administrators, and businessmen seem to have been at the front of the queue in the allocation of private land rights. They too have built new houses and put fences round their estates and gained status and influence through their conspicuous control of a large personal territory. They too are unlikely to see their territory as something they hold in trust for future generations, to be actively looked after, and carefully cultivated and protected. There is much evidence already of speculative land sales for private housing, on sites without adequate infrastructure and on good agricultural land, where an effective land-use planning system would never allow such development. However, in such Eastern European countries as Macedonia and Bulgaria, as well as in Zimbabwe, an effective land use planning system is simply not present in these countries, and so good land is spoiled and bad suburbs are created.

The solution is easy enough to propose, but almost impossible to impose. New nations have to find their own way, though they would be wise to look around them for examples of land tenure systems, which work, or at least have survived the test of a few generations of operation. The four stages set out at the beginning of this chapter must be accompanied by a fifth, which is to establish a system of central land use planning, supported by robust legislation, with implementation delegated down to local level. Indeed all five stages *should* occur almost simultaneously, or alternatively the second – legislation, which defines the nature and extent of the rights, which citizens can hold over land and buildings – should very clearly set out those uses of land or buildings which can be freely exercised and those for which the permission of higher authority is required. Unfortunately, the genie has been let out of the bottle already, and it will be fiendishly difficult to get him back in again. Land rights have been created and distributed prematurely, in both Eastern Europe and Southern Africa, in advance of the necessary laws to restrain unsocial uses. Furthermore, because the new owners have acquired status as well as assets, influence along with personal territory, they will be well equipped to resist the imposition of the necessary checks retrospectively. At both

---

[4] Virtually every rural village in Southern Africa has experienced a loss of young members, especially young men, to the towns. More recently, many are losing both young and middle-aged men and women through AIDS.

local and national level they will be able to exploit the deference which 'goes with the territory', perhaps for many years.

Of course, if the view typified by the quotation from Hobhouse given above is correct, property ownership will inevitably have a moderating effect on the selfish, aggressive, assertive personalities of the new proprietors: the new owners will become more rational, more responsible, more confident in their self-image, and therefore less exploitative in the use of what they own. They will not need to perform the human equivalents of the defensive territorial displays performed instinctively by animals and birds in nature because their territories will not be challenged by intruders, being protected by laws of trespass and guaranteed by a state register of titles. Moreover, in time they will find that they gain more status and respect from the fairness, wisdom, and foresight of their proprietorial actions than from the flamboyant or ruthless display of their landed wealth. Thus does the Mafia of one age become the aristocracy of the next. In the meantime, there will continue to be turbulence in the transitional process.

## References

Altman, I., 1975. *The Environment and Social Behaviour: Privacy, Personal Space, Territory, Crowding*, Brooks/Cole, Monterey, CA.

Ardrey, R., 1969. *The Territorial Imperative: a Personal Enquiry into the Animal Origins of Property and Nations*, Collins, London.

Ardrey, R., 1970. *The Social Contract: a Personal Enquiry into the Evolutionary Sources of Order and Disorder*, Collins, London.

Arnison, C.J., 1984. Masters and tenants: leasehold tenure as a mechanism for moral governance. In: Chiddick, D. & Millington, A. (eds.), *Land Management: New Directions*, Routledge, London.

Arnison, C.J., 1990. Property and personal development. In: Grover, R. (ed.), *Land and Property Development: New Directions*, Chapman Hall, London.

Beaglehole, E., 1931. *Property: a Study in Social Psychology*, Allen and Unwin, London.

Clay, T., 1624. *A Chorological Discourse of the Well Ordering, disposing, and governing of an Honourable Estate or Revenue*, London.

Csaki, C., 1994. Where is agriculture heading in Central and Eastern Europe? Presidential Address to the *XXIIth International Congress of Agricultural Economists*, Harare, Zimbabwe.

Davis, K., 1948. *Human Society*, Macmillan, New York.

Dickinson, H.T., 1977. *Liberty and Property: Political Ideology in Eighteenth Century Britain*, Methuen, London.

Etkin, W., 1964. *Social Behaviour and Organisation among Vertebrates*, Chicago University Press, Chicago.

Hobhouse, L., 1913. The historical evolution of property, in fact and in idea. In: Gore, C. (ed.), *Property: Its Duties and Rights*, Macmillan, London.

Hollowell, P.G. (ed.), 1982. *Property and Social Relations*, Heinemann, London.

Home, R., *Turning Fish Soup into Fish: Towards New Land and Property Markets in Bulgaria*, RICS Research Paper Series, **1**(9).

Irving, P.G. and Goldstein, S.R., 1990. Effect of home-field advantage on peak performance of baseball pitchers, *Journal of Sport Behaviour*, **13**(1), 23-27.

Kalabamu, F.T., 2000. Land tenure and management reforms in East and Southern Africa – the case of Botswana, *Land Use Policy*, **17**, 305-319.

Malmberg, T., 1980. *Human Territoriality*, Mouton de Gruyter, New York.

Newby, H., 1977. *The Deferential Worker: a Study of Farm Workers in East Anglia*, Allen Lane, London.

Reeve, A., 1986. *Property*, Macmillan, London.

Schlatter, R., 1951. *Private Property: the History of an Idea*, Rutgers, New Brunswick.

Scott, J., 1982. *The Upper Classes: Property and Privilege in Britain*, Macmillan, London.

Thiesenhusen, W.C., 1994. *Landed Property in Capitalist and Socialist Countries: Nature of the Transition in the Russian Case*, Land Tenure Center, University of Madison-Wisconsin, USA.

Trasler, G., 1982. The psychology of ownership and possessiveness. In: Hollowell, P.G. (ed.), *Property and Social Relations*, Heinemann, London.

Chapter 3

# Changes in Land Use and their Implications upon Coastal Regions: The Case of Grange-over-Sands, Northwest England

Robert Dixon-Gough

## Introduction

Grange-over-Sands is a tourist resort located in the North West of England, situated on the northern shoreline of Morecambe Bay. It has a southerly aspect and enjoyed a reputation as a popular resort for middle-class holidaymakers during the nineteenth and early twentieth centuries. As in the case of many coastal resorts throughout England and Wales, there has been a settlement on this site for many years and, indeed, evidence of a harbour. Although once situated directly on the shoreline of Morecambe Bay, the construction of an embankment carrying the Lancaster and Ulverston Railway effectively separated the resort from the sea, leaving it to develop behind this barrier.

During the early part of the twentieth century, the shoreline of the resort had a tendency towards fine silt rather than courser sand, in part influenced by deposition from the rivers Winster and Kent. The River Kent, in particular, has a reputation of cyclic channel changes within its estuary, which is responsible for the accretion and erosion of material either to form or from estuarine salt marshes. Whilst there is considerable historical evidence of channel shifts and the subsequent effects and implications of those shifts, there is no evidence to imply that salt marshes existed close to the foreshore of Grange-over-Sands. The main areas of salt marsh have either been on the eastern side of the Kent estuary, between Arnside and Silverdale, or to the south of Grange-over-Sands around the Cartmel peninsula. However, since the mid 1980s there has been a considerable accretion of material along the foreshore of Grange-over-Sands leading, through the introduction of the pioneering species *Spartina Anglica*, to the establishment of a salt marsh, which is now being used to graze sheep and cattle. Conversely, since the mid-1990s there has been the virtual destruction of the salt marshes on the eastern side of the estuary caused by the erosional effects of the deepwater channel of the River Kent. There is no historical precedence for the establishment of salt marshes around Grange-over-Sands, although there is evidence of cyclic erosion of the Silverdale salt marshes.

The most likely cause for this change can be derived from an analysis of changes in land management practices both along the shoreline of the Kent estuary and throughout its catchment area. The general location of Grange-over-Sands is illustrated in Figure 3.1. Whilst the major contributing factor is likely to be the construction of the railway and its infrastructure, together with the large areas of land claim associated with its construction, other events have also had a contributory effect. These include land drainage, such as the drainage of 'mosses' or bogs, improvements and modifications to the watercourses, and changes in land management practices. All in their own way are relatively minor but when taken together, represent significant implications to the nature of the catchment and the estuarine environment and, as a result, the changes that are taking place around Grange-over-Sands. Furthermore, it is interesting to speculate whether those changes to the foreshore of Grange-over-Sands are permanent or whether they are likely to change in the near or distant future. It is possibly that what might happen in Grange-over-Sands, will mirror the situation at Silverside, where a series of cycles associated with the migration of the deepwater channel of the Kent over the last 100 years has resulted in the erosion and re-establishment of the foreshore salt marshes.

One of the most interesting aspects of this study has been the way in which the perceptions of the inhabitants of Grange-over-Sands have changed through the last decade. During the early part of the establishment of the salt marsh, there was a great deal of pessimism concerning the future of the resort. Now that the marsh is fully established, this pessimism has turned to optimism as the resort is currently experiencing a mini-boom in holidaymakers. One of the endearing characteristics of Grange-over-Sands is that it never became truly commercialised during the post-war years, as did many similar resorts and retains its Edwardian elegance. This is helped, in part, by the effect of the railway, which effectively cuts the resort off from the sea and allows it to exist in a comfortable manner as Grange-over-Sands (without the sand).

## The coastal resorts of Morecambe Bay

Although coastal resorts had been part of the British landscape and social calendar since the middle part of the eighteenth century (Walvin, 1978), it was the relative affluence of the industrial revolution that permitted lower middle and working class members of society to enjoy similar delights (Dixon-Gough, 2000). The rapid expansion of the railways during the 1830s to 1850s generated a significant social revolution and this provided a catalyst for the development of many coastal resorts (Dixon-Gough, 1998; 1999).

The construction of the Lancaster to Ulverston railway had a profound affect upon settlements along the coastline of Morecambe Bay (see Figure 3.2).

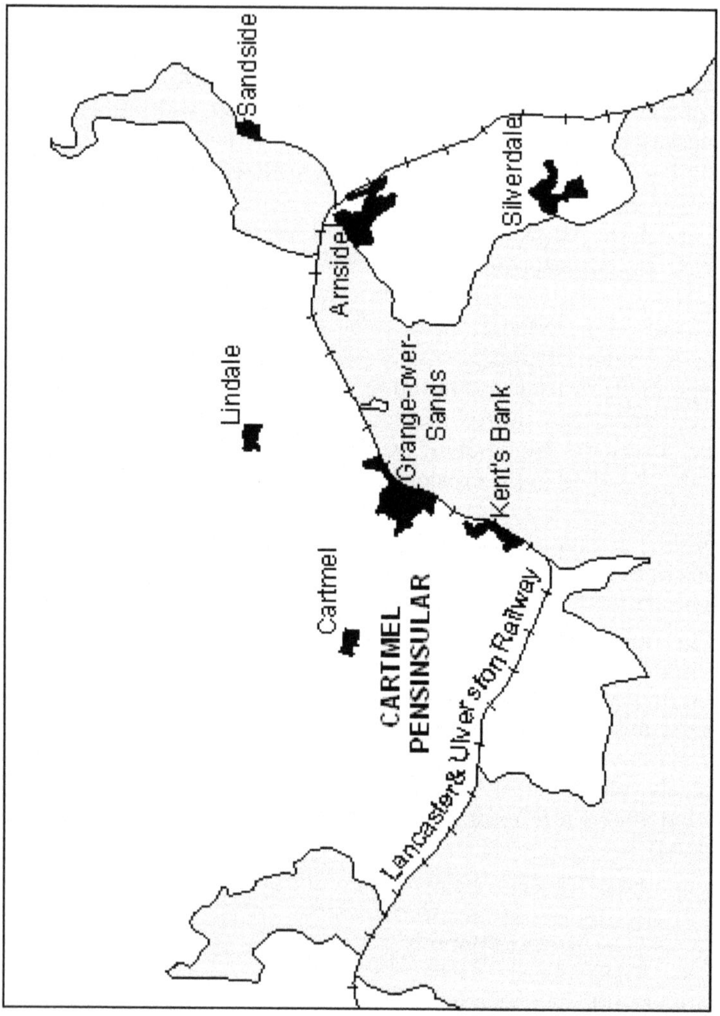


**Figure 3.1 General location of Grange-over-Sands**

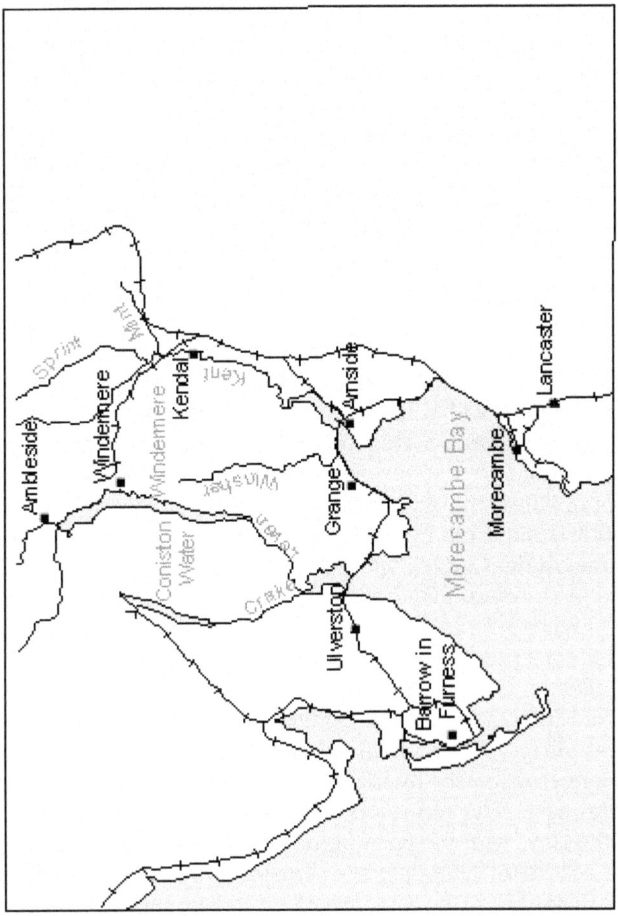


**Figure 3.2 The geography of the northern part of Morecambe Bay**

To some extent, the present resorts were all active before the coming of the railway and many of the cottages and farmsteads had been modified to cater for the needs of the very few summer visitors. The railway made these resorts more readily accessible, and although they grew to cater for the increased need, the influence of the original settlements is still evident in the current layouts.

Prior to the 'railway age' several settlements around Morecambe Bay enjoyed a reputation as watering places, the majority being along the shores of the Kent Estuary at Grange, Milnthorpe, Sandside, Arnside, and Silverdale. The settlements were all small. For example, in the census of 1851 the population of Silverdale was 240, and that of Arnside 167. Similarly, the total population of Grange (the name Grange-over-Sands being a later 'modification') and neighbouring hamlets was 260.

Much of the evidence relating to the early layout of Grange has been destroyed by the construction of the railway and railway embankments. However, Dickinson (1980) considered that there is sufficient evidence to suggest that there was a pool from which local shipping could sail, at least at high water. There originally seems to have been a ridge of rock separating the sea from a narrow channel, much of which is now covered by the present pond in the ornamental gardens, and a pool below the present position of the Grange Hotel. This is borne out by examination of the First Edition, Ordnance Survey 1:10,560 Country Series Plan, for which the original survey was completed before the arrival of the railway. It shows the inlet and, where the Grange Hotel is currently situated, a farm called Windy Harbour (Figure 3.3).

Before the construction of the railway, the majority of visitors to Grange would have travelled on one of two routes, either across the Bay from Hest Bank to Kents Bank or via Milnthorpe, Levens Bridge and Lindale. The first route, over the sands, was more direct, faster, although not without risk, whilst the second involved much longer distances owing to the necessity to cross the River Kent at the head of the estuary. The influence of the railway was initially felt along the western part of the Bay, with the opening of the Lancaster and Carlisle Railway in 1846. This gave access to Silverdale and Arnside from railway stations located at Carnforth and Milnthorpe, with many travelling to Milnthorpe on cheap excursion tickets from Lancaster. Despite this, the Cartmel peninsula remained relatively isolated until the opening of the Ulverston and Lancaster Railway in 1857. Stations were built at Grange and Kents Bank when the line was initially opened and a temporary station opened at Arnside the following year.

Although the railway had originally been constructed for mineral traffic, the directors of the company were aware of the holiday potential of the line (Gilpin, 1997). In conjunction with other railway companies, excursion trains were run to places around Morecambe Bay. One of the major drawbacks of the settlements around the Bay was their lack of accommodation, but privately built accommodation began to increase within a very short time.

The construction of the railway effectively cut off the old settlement of Grange from the sea by a high embankment. All new developments were, therefore, restricted to the limestone slopes away from the sea.

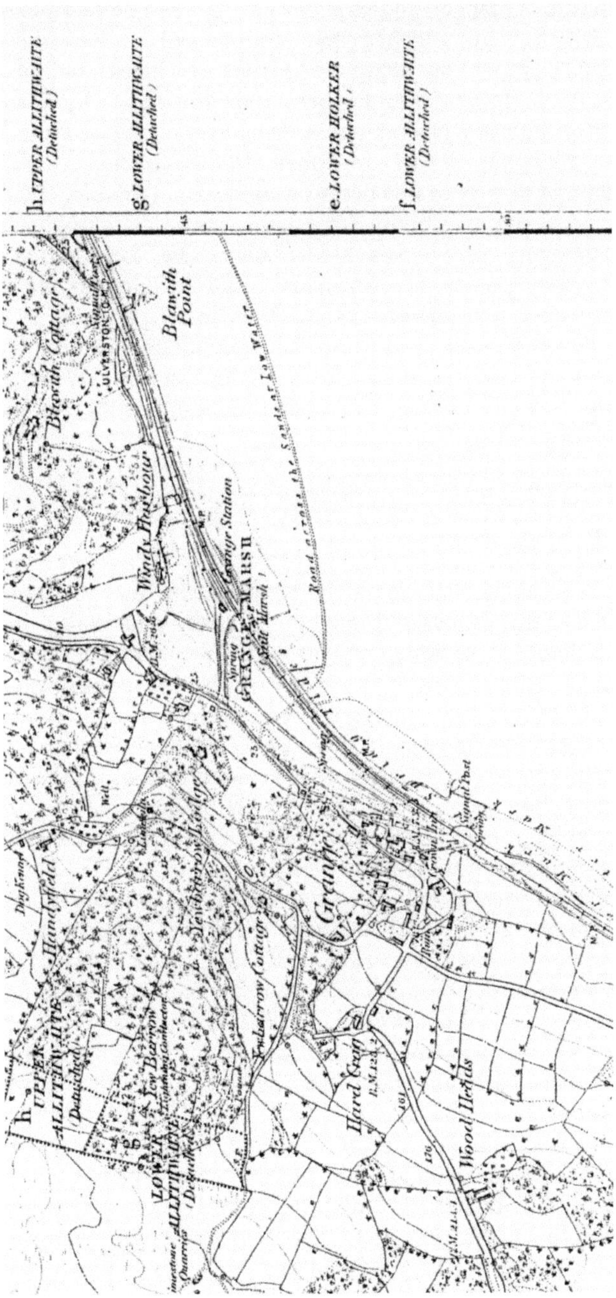


**Figure 3.3 Extract from the 1st Edition Ordnance Survey Country Series (Lancaster) 1:10,560 Plan** showing the extent of Grange and its harbour, surveyed prior to the construction of the railway. The map is dated 1851.
*Source*: Reproduced by kind permission of the Cumbrian Record Office, Kendal Branch. Crown Copyright Reserved.

The growth was unplanned and largely followed the line of the old paths and roads, many of which were widened and straightened, and as the expansion increased, new roads were laid out along the hillsides and houses built along those.

As with many coastal resort of this period, the town grew by housing people who had retired from industry or commerce, from service in the armed forces, or from a career in the Empire (Dixon-Gough, 2002). Many of the amenities associated with popular resorts of this period were absent. Between 1902 and 1904, an improved sewage system was built with a mains sewer laid along the sea side of the railway embankment, which was hidden under the construction of a new promenade. The resort possessed two piers, used as landing stages for the Morecambe boats rather than attractions in their own right until 1910 when a movement of the River Winster away from the shoreline reduced the water depth. The resort was also affected by the vagaries of the Kent and Winster channels, which from time-to-time exposed on the foreshore limestone outcrops, shingle, and sand.

**Land reclamation**

One of the most striking impacts upon the Kent estuary was the extent of land reclamation. Unlike many estuaries throughout England and Wales, many of the land reclamation schemes were not commenced until the eighteenth century and, as a result, much of the estuary is still in a state of physical adjustment. The Kent estuary was fringed by extensive areas of peat, partly covered by raised beach deposits, known locally as 'mosses' (Gray, 1972). Until the construction of the railway, land reclamation was piecemeal and has largely involved the enclosure of small embayments as they become silted up. Within the shores of Morecambe Bay, approximately 1,300 ha of land have been claimed, of which approximately 400 ha have been claimed directly for agriculture.

In 1851, the Ulverston and Lancaster Railway Company (UandLR) was proposed, which involved the canalisation of the River Winster with the Kent estuary being crossed by means of a viaduct. Because of the nature of the estuary, the intention was to construct the railway on an embankment along the foreshore and across the mosses and saltmarshes that already existed. However, an integral part of the programme involved land reclamation, although this was confined to the marshy areas of Holker, Meathop and Silverdale situated on the landward side of the railway embankment (Gooderson, 1970). The UandLR was exceptional amongst rural railways insomuch that it was able to recoup some of its losses through the sales of reclaimed land at Cark, Meathop, and Silverdale.

The Ulveston and Lancaster Railway (UandLR) Act became law in July 1851 and work began on the first section in May 1852, the engineering works across Winster Bay between Blawith Point and Meathop Point. It had been the original intention of the UandLR to cross the River Winster by constructing a bridge or viaduct. This was amended on the possible suggestion of John Brogden (the railway contractor, who also had interests in land development) that the sands on Winster Bay be reclaimed and developed as farmland (Gilpin, 1997). As a result, the River Winster was canalised to Meathop Point, where it passed through a sluice under the railway embankment. This led to a perceived emphasis upon land reclamation rather than railway construction, which resulted in criticisms being directed towards the UandLR and the Brogdens in that they were primarily a land company rather than a railway company. This was confirmed in the UandLR Act of 1858, which allowed both for extra capital to be raised and for the company to be sold to the Furness Railway, but also for the restructuring of the company as the Ulveston Land Company.

Other important consequences to the Kent estuary included the construction of the Arnside to Sandside railway, which protected of the high level saltmarshes and mosslands to the east of the Kent estuary (Arnside Moss). There were also further, minor, reclamations on the Cartmel peninsula between Cark and the viaduct over the River Leven.

The principal direct impact of the railway was the enclosure and reclamation of several hundreds of acres of land to the south-east of Meathop village. Similarly, a large area of the Winster valley was reclaimed and the course of the Winster canalised to increase river flow and drainage. This effectively fixed the boundary between Lancashire and Westmorland, ensuring that Holme Island no longer alternated between the two counties, as the channel of the Winster meandered across the channel. Holme Island, which was purchased by Alexander Brogden, was linked to the mainland through the construction of a causeway to the island.

The result of these intrusions, embankments and constructions within the estuary had significant physical effects. One of the most pronounced was the reduction of the tidal cubature within the estuary and a consequential reduction in the velocity of the ebb-tide. This alone has encouraged the deposition of sediment from the suspended load and has resulted in an increased rate of accretion (Inglis and Kestner, 1958; Kestner, 1956, 1970; Kestner and Inglis, 1956). It is also possible that through the construction of training walls and embankments, the fluctuations of the channels might be stabilised, thereby reducing the effects of erosion. The increased rates of accretion had a direct impact as the development of the salt marshes within the area increased.

One of the earliest consequences of this was that ships using Port Milnthorpe effectively ceased by 1855. Although the railway had not, at this time, been opened, most of the construction work had been completed and these works might have led to the silting up of the port. Arnside pier was eventually constructed during the late nineteenth century because ships could no longer travel up the estuary because of increased silting. The silting resulted in an increased growth or extension of the existing salt marshes (Theaker, 1981).

## The River Kent catchment

Any water catchment area is under a constant state of change, which has implications upon the whole catchment and, in particular, those downstream riparian and estuarine zones. The hydrological characteristics of a catchment may determine the ways in which the land is used, the land use, and the techniques of land management, which in turn could significantly influence the hydrological response of the catchment (Burt et al., 1993). Human impacts and climatic changes can modify the pattern of runoff within a catchment. In the case of rural headwaters, this is determined by factors such as the vegetation cover and in those areas transitioning between pastoral to mixed arable farming, where the removal of grass can result in greatly increased rates of wash erosion. Certain locations may be particularly sensitive to changes in land use and land drainage, especially those areas with relatively short and fast running rivers.

Prior to the programmes of land improvement, which in Kent catchment effectively date from the early parts of the nineteenth century, the land was infertile owing to lack of drainage and consisted largely of:

> extensive morasses, swamps, pits, stagnant sheets of water, obstructed streams, and much under drained, or at any rate, ineffectively drained land (Stockdale, 1872).

Higher up the slopes of the fells, the land had no use for cultivation so was largely left for stock to wander (Dickinson, 1980).

One of the most fundamental changes to this landscape occurred as the result of the Parliamentary Enclosure Awards, which took place within the Kent catchment area between 1836 and 1874. Although the main purpose of the Acts and Awards was separate common or wasteland that had a shared ownership into enclosures having individual ownership and responsibility, thereby improving agricultural efficiency, the mechanics of process had significant implications on land drainage and upon the hydrological cycles of upland catchment areas (see Jenkins, 2002; Williamson, 2000; Yelling, 1977). In addition to the Enclosure Awards, Drainage Acts and Awards had a similar function in ‘improving’ intertidal areas. A significant element of both the Enclosure and Drainage Awards was concerned with the construction of roads and bridges, the drainage of mosses and boggy land, and the straightening of river courses. These had significant impacts upon the physical characteristics of the areas and, in particular, the rate of run-off within the catchment area.

The land management changes and effects upon the drainage within the catchment area is summarised in Table 3.1. Although some of the effects are identified in Table 3.1, the overall implications on the hydrological cycle of the water catchment has significant impacts upon the dynamics and nature of the river flow. The consequences of these actions were not immediately obvious and, in many instances, were blamed upon more obvious actions, such as the construction of the railway embankment and viaducts.

One of the earliest effects of the changes to the hydrology of the River Kent came during the 1840s, when the owners of water-powered mills in Kendal and up-stream of the town noticed increased seasonal variations in the flow of the river. This resulted in a meeting of the mill owners, reported in the Westmorland Gazette of 17$^{th}$ August 1844, to canvas support for the construction of a regulating reservoir at Kentmere:

> We understand that a project of considerable importance has been planned by owners of water mills along the course of the river Kent, with a view of obviating the great inconvenience which has been experienced during the past and other years from the want of sufficient water to drive the mills. The project proposal is to form a vast embankment at Kentmere, so that the waters accumulated in winter may furnish a better supply in droughty seasons that has been hitherto enjoyed; and it is alleged that the scheme will be comparatively easy to execute, as an embankment will be required on only one, the south, side of the projected artificial lake, the other sides of the basin being formed naturally by the hills … it is intended to apply to parliament for powers to carry it into execution.

The following edition of the Westmorland Gazette, on the 24th August 1844 reported that:

> The proposed embankment at Kentmere, noticed in our last paper, is not a mere fanciful project, but has assumed a definite and promising shape, a meeting of influential gentlemen, whose practical judgement the public will be quite willing to rely on, having been held on Thursday … For several years, and especially during the last summer, great difficulties have been experienced by mill-owners on the Kent front, the scarcity of water, so much so, that one manufacturer has estimated his loss of orders from this cause as £1,000 in one year alone. This difficulty has been aggravated by the progress of agricultural improvements, in draining the land, etc., which, however beneficial in itself, has had the effect of depriving the mill-owners of a large quantity of water power, the waters going down more rapidly, and of course sooner running away. Under these circumstances it has been determined, as we intimated last week, to form a large Reservoir, at Kentmere, by which might be stored up a sufficient supply of water during the season of flood, and thus obviate the inconveniences both of scarcity in dry seasons, and an over abundant supply during thaws in winter, or, as one of the speakers remarked, both of lack water and back water.

In addition to this noticeable affect of the land improvements of the Enclosure Awards, it is likely that other effects might have been noticed and blamed upon other causes. For example, Theaker (1981) cites the example of the Port of Milnthorpe, which effectively ceased by 1855, blaming the construction of the railways although it had not, at this time, been opened. However, most of the infrastructure had been completed and Theaker (1981) considered that these works might have led to the silting up of the port. However, in the light of the evidence of the mill-owners in Kendal and of the drainage work that had been undertaken on the opposite side of the estuary at Heversham, there is a more likely explanation. The changes to the hydrological characteristics of the River Kent resulting in a more seasonal flow of water, which when combined with the loss of intertidal areas upon which fine silt had previously been accreted, were the most likely cause of increased siltation at the Port of Milnthorpe. Furthermore, this change in the hydrological cycle is accumulative and increases in both time and location, affecting areas further down the estuary (Kestner and Inglis, 1956).

In addition to the effects of the Enclosure Awards, other forms of land improvements and variations in land management have been practised during the nineteenth and twentieth centuries. These have all had implications to the hydrological characteristics to the River Kent and its estuary. For example, the carrying capacity of a water catchment area is an estimate of the amount and type of vegetation that is grown, together with the number of animals that can be supported. This is normally expressed in terms of an annual or monthly grazing cycle, which in practical terms is defined by the amount of vegetation needed to support a particular animal. Wilson and Harrington (1984) provided the example of 1 head of cattle = 8 sheep = 11 goats = 133 rabbits. However, this ratio does not take into consideration the amount of erosional impact that an animal has upon the landscape. This is a product of both the amount of vegetation consumed by the animal together with the effect of the hooves impacting upon the land. Thus grazing intensity should be based upon the number of animals grazing the land, expressed in animal/ha. Of greatest interest in respect to the erosional impact is the maximum grazing intensity, since it is at this level at which most vegetation is removed and damage to the land inflicted through trampling. Also of interest, is whether the land is permitted to recover during periods of less intensive grazing when, for example, sheep are brought down from the hills for sale at market or in sheep fairs?

In areas such as the River Kent catchment, the periods of lower stock intensity coincide with the winter months when the climate is too severe for vegetation to recolonise the bare soil. On the peat moorlands typical of the Kent catchment, sheep scars are formed at summer grazing intensities of a sheep/2 ha (Evans, 1997).

There has also been a strong positive correlation between the increase in sheep numbers since the late 1940s and the increased rate of erosion (Evans, 1997). Animals have grazed these upland areas for centuries, normally during the summer months. Since the late eighteenth century, the main grazing animal has been sheep and the upland areas stocked for virtually the whole year. These sheep were initially shepherded but have been allowed to roam freely for many years. The numbers have increased steadily since 1875, the year of the first agricultural census, and continued to increase up to the beginning of the foot and mouth outbreak in 2001. Taking the late 1940s as a baseline, the number of sheep have increased by between 150 and 200%, initially in the Peak District but more latterly (1980s) in the Lake District (Evans, 1997). Evans (1993) provides extensive evidence of the effects of erosion ascribed to sheep. Grazing densities of about 0.5 sheep per ha was sufficient to cause a decline in the heather moorland, and to then initiate erosion in the form of sheep scars within the better-quality acid grasslands. Increased grazing pressures can lead to increased runoff (Evans, 1990a), leading to increased sediment in streams as well as channel erosion.

Human interactions can also affect the sensitivity of the land to erosion through trampling vegetation and exposing soil. Many ancient trackways and footpaths have become deeply eroded. In those areas where tracks have been opened up for walkers, farmers, stalkers, etc., these may act as catalysts for erosion (Watson, 1985).

**Table 3.1 Some comments from Enclosure Awards of the Kent catchment resulting in modified river flows**

| Enclosure Award | Location | Action | Effects |
|---|---|---|---|
| Kentmere Enclosure Award. Date of Act 1836. | Valley of the River Kent | … and cut away in the formation of a new and improved River Course through the said High Field … also the catchwater drains through and out of the High Fields shall be hereafter maintained repaired kept up cleansed out and preserved in a substantial and proper manner ... | Straightening of the River Kent to reduce the length of the river and to increase the 'efficiency' of flow. Drainage of the land, which acts, during the winter months, as a water store to naturally regulate the flow of the river. Both have resulted in improved drainage but more instantaneous run-off contributing to 'flashier' river levels during the rainy period and low water levels during the dry periods. |
| Green Quarter Pasture in the Township of Kentmere in the Parish of Kendal. Date of Act 1853. | Valley of the River Kent | .... And I do order and direct that in getting such peats they shall not dig or leave any holes in which the water way be lodged or deposited but shall take due care to cause the water to run off … and to this end they shall cause proper drains or outflow ditches to be provided so that all water may regularly flow away from the deepest part of such mosses … | Drainage of the mosses, which act, during the winter months, as water stores to naturally regulate the flow of the river. This resulted in improved drainage but more rapid run-off contributing to 'flashier' river levels during the rainy period and low water levels during the dry periods. |
| Longsleddale (Sadgill Common or High Fell). Date of Act 1865 | Valley of the River Sprint, a tributary of the River Kent | ... for the purpose of getting stone for repairing the embankment at Goatscar Thwaite in order to prevent the sand from flowing down the river … | Modification of the flow of the river and of the amount of sediment flowing down the river. |

In areas such as the Lake District National Park, where the number of walkers has increased enormously throughout the last 60 years, this has had a great effect upon the erosion of paths. Furthermore, runoff channelled from roadside drains, cattle and farm tracks onto slopes can also generate extensive gullying and erosion (Duck, 1985).

Evans (1990b) provides an extensive review of diary sources that indicate that the sensitivity of the landscape to erosion was not unusual during the eighteenth and nineteenth centuries. One of the main differences to the current situation is that more land is sensitive to erosion than in earlier times since more is under arable cultivation and, furthermore, much of the arable land brought into production over the past 250 years is still vulnerable to erosion. This is certainly true in the upper reaches of the Kent estuary, where the silt-laden run-off combined with the high levels of nitrates has had an impact upon the accretion of salt marshes, particularly in the lower reaches of the estuary and around the Cartmel peninsula.

Agricultural policy changes, particularly those directed at reducing food production and the restocking of regions of upland England following the foot and mouth epidemic, will result in wide-scale changes in the rural landscape (Dessylas, 1990; Meeus et al., 1990; Cherrill, and Lane, 1995). Many changes will take the form of modifications to areas and their land cover, particularly in the case of those areas such as the northern uplands of England, where the overall population of sheep has been culled and restocking has been at a much reduced level. Some indication of the level of concern of the farming community in this region has been from interviews conducted in the region during January 2002 (Buick, 2002). The concern is that upland livestock culled during the foot and mouth outbreak will be restocked at the rate of 3 to every 10 culled.

Other implications of changing agricultural practices includes the transition from pastoral to arable farming, and the release of farm land for rural and peri-urban development, both of which have severe implications upon runoff into the water catchment and to the hydrological characteristics of the River Kent. A wide range of factors, such as existing land use, land and property prices, soil quality, aesthetic appeal, hydrology, pollution, and wildlife interests are all potential determinants of the location and nature of changes in land cover (Whitby, 1992).

Finally, there are the implications of climate change, not in a dramatic sense but in the changing patterns of rainfall that will have a significant effect upon the hydrological characteristics of a water catchment such as that of the River Kent. There is good statistical evidence of short-term precipitation trends in England and Wales since the 1970s indicating an increase in the frequency of extreme dry summers, wetter winters and springs, together with periods of extreme precipitation (Limbrick, et al. 2000). Robson et al. (1998) suggests that whilst there has been no significant trend in national flood frequency or magnitude since 1870, flood magnitudes and frequencies may be linked to both changing trends in the fall of precipitation within catchments and to changes in surface runoff primarily the product of vegetation, farming, and urbanisation (Loaiciga et al., 1998). This has been verified by the author in interviews conducted with farmers, who emphasise the 'flashy' nature of streams, particularly in the upper reaches of the catchment of the River Kent. These factors have an undoubted effect upon the hydrological cycle of the River Kent and play an important role in the movements of the deepwater channels in the estuary.

## Culmination of effects in the Kent estuary

During the first half of the nineteenth century, the coastal settlements were very accessible to the sea and custom houses were built at Arnside, Milnthorpe, and Sandside. Since there had been no attempt made to train the channel of the Kent, it was allowed to meander freely across the estuary. The ebb tide was also very strong, which meant that until the 1850s, there was little tendency for the estuary to silt-up and very little formation of saltmarshes.

In the latter part of the nineteenth century, the potential for land claim for agricultural purposes in the estuary was recognised. The most significant of these has been identified by Gray and Adam (1974). Probably the most significant factor to affect the morphology of the estuary was the construction of the Lancaster to Ulveston Railway. This altered the outline of the estuary to a far greater extent than any other single activity, leading to extensive areas of land reclamation (Gray and Adam, 1974).

A key factor in the accretion of saltmarsh, the precursor of all reclamation activity, is the stabilisation of the low water channels. Erosion along the east bank of the Kent estuary occurred particularly during the seventeenth and eighteenth centuries, as far down as Hest Bank. Lucas (1744) stated that:

> The Kent some time ago washed away 200 acres of marsh and exposed the site of shipyards of 200 years before ... [It] seldom continues long in one place. Sometimes by the violence of a rough tide, its channel will be removed at once, a considerable way from where it ran before ... when the river draws near to either side of the sands, it becomes a troublesome neighbour and generally destroys all or most of their marshes [parenthesis added].

The cyclic nature of deepwater channels in estuaries is common, and is generally ascribed to a combination of factors such as the state of tides and the hydrological cycle of the rivers. It is normally assumed that the development of saltmarsh in estuaries has been subject to the changing position of the low-water channels, which may produce alternating phases of erosion and accretion in any particular location. One such example in the Kent estuary is Silverdale, which provides an example of the cyclic nature of channel changes. Details of such changes can be derived from a combination of old estate maps, Ordnance Survey map and plans, aerial photographs, and both documented and anecdotal accounts. During the 1840s, Silverdale was a popular bathing resort but this activity declined during the 1850s as the foreshore became progressively muddier (Harris, 1962) and by 1861, a small area of saltmarsh had developed (Waugh, 1861). The Ordnance Survey plan of 1893 showed an extensive area of saltmarsh at Silverdale (approximately 127 ha). This rate of accretion is associated with the lateral movement of the Kent low-water channel as it moved to a position close to and parallel with the Grange shore. It remained in approximately this position until the early part of the twentieth century, when it moved across towards the Silverdale shore, destroying most of the saltmarsh. 'Silverdale is changing and the old marsh is almost gone' (Ashton, 1909). By 1915, the channel had returned to the Grange shore (Pape, 1947) and the saltmarsh continued to develop through most of the twentieth century. Gray (1972) reported that the saltmarsh at Silverdale increased in size some 135 ha between 1946 and 1967. During a period from the late 1970s, the Silverdale marshes have been extensively eroded (Gray and Scott, 1987) and the erosion continues (Pringle, 1995; 2002).

Kestner and Inglis (1956) observed that in the case of rapid and cyclic channel movements in estuaries, a detailed study of changes in the smaller upstream stretches of the river might provide evidence for this process. This is particularly so in those estuaries where deposition from sediment was an important factor. An examination of upstream activities might supply information regarding the factors that control the accretion and erosion of the rapidly changing banks and channels along the margins of the estuary. This gives emphasis to the implication of land management changes throughout the catchment area.

The situation is even more complicated in the case of the Kent estuary, which has been subjected to many changes. When land has been reclaimed and enclosed by an embankment, accretion is likely to be hastened both by reducing the total volume of flow, and the velocity of local flow. Initially this will have little effect on the flood tide but will slow down the ebb tide and, as a result, increase the rate of accretion. This will cause further reductions in velocity with further deposition. This would certainly account for the siltation caused at Port Milnthorpe and also the likely impact upon the foreshore at Grange-over-Sands caused by the extension of the training wall of the River Winster into the centre of the Kent estuary.

It would seem likely and predictable that with the changes to the catchment of the River Kent and changes to the morphology of the estuary, that the accretion of fine deposits instrumental in the formation of the salt marsh will remain a feature at Grange, and in many other areas in the lower part of the estuary. It is, therefore, more than possible that the situation encountered at Silverdale will repeat itself at Grange-over-Sands and the cyclic erosion and accretion of the salt marsh. It is unlikely that this will seriously affect the tourist potential of the town and, because it is protected by the promenade and the railway embankment, any erosion – even in its most extreme case – will leave the town no worse off than it was when it was first developed.

## Conclusion

There is a relationship between historical land use and land management practices and the cycle of physical and 'natural' events. Grange-over-Sands, the subject of this study sits on the estuary of the River Kent, which as had a long reputation for cyclic movements. The evolution of the coastal resort within the context of the land management changes throughout the catchment of the River Kent have a distinct correlation between related causes and events that have, and continue to, take place in the estuary.

One of the most important of these has been the accretion of salt marsh along the foreshore of Grange-over-Sands, where there has been no historic evidence of such accretion in the past. This can be accounted for by land and drainage improvements within the catchment area together with significant changes to the shoreline such as land reclamation, land drainage, and the construction of the railway. These events have combined to alter the hydrological characteristics of the river and of the hydrological cycle of the estuary, which has had the effect of modifying the locations in which fine silt is accreted to the lower parts of the estuary. This was initially encountered at Arnside and Silverdale, from the 1850s, but more recently at Grange-over-Sands.

Although the characteristics of the resort have changed from having a beach to a salt marsh, the resort remains as popular as ever. The changes to the location have been very significant but Grange-over-Sands, protected from the shoreline by the railways, remains effectively untouched.

## References

Ashton, W., 1909. *The Battle of Land and Sea*, Heywood and Sons, Manchester.

Buick, P. (2002) Interview conducted by Robert Dixon-Gough at Bouth, Cumbria, 17th January, 2002.

Burt, T.P., Heathwaite, A.L. and Trudgill, S.T., 1993. Catchment sensitivity to land use controls. In: Thomas, D.S.G. and Allison, R.J. (eds.), *Landscape Sensitivity*, John Wiley and Sons, Ltd, Chichester.

Cherrill, A. and Lane, M., 1995. The survey and prediction of land cover using an environmental land classification, *Applied Geography*, **15**, 69-85.

Dessylas, M.B., 1990. The adoption of the social and structural policy of the EEC to the changed market situation: the protection of the countryside, *Landscape and Urban Planning*, **18**, 197-201.

Dickinson, J.C., 1980. *The Land of Cartmel: a History*, Titus Wilson and Sons, Kendal.

Dixon-Gough, R.W., 1998. An analysis of the interaction between tourism and nature within the coastal zone, with specific reference to Morecambe Bay, 28th International Symposium of the European Faculty of Land Use and Development, Kingston University.

Dixon-Gough, R.W., 1999. The railway town: a case study of urban sustainable development. In: R.W. Dixon-Gough (ed.), *Land Reform and Sustainable Development*, Ashgate Publishing Ltd, Aldershot.

Dixon-Gough, R.W., 2000. An analysis of the planning and regulatory instruments of Great Britain with particular reference to the development of the British coastal resort, *Social Strategies*, **31**, 171-178.

Dixon-Gough, R.W., 2002. Cause and effect: the influence of land management upon the coastal zone. A case study of Morecambe Bay, North-west England, *Social Strategies*, **36**, 145-162.

Duck, R.W., 1985. The effect of road construction on sediment deposits in Loch Earn, Scotland, *Earth Surface Processes and Landforms*, **10**, 401-406.

Evans, R., 1990a. Erosion studies in the Dark Peak, *Proceedings of the North of England Soil Discussion Group*, **24**, 39-61.

Evans, R., 1990b. Soil erosion: its impact on the English and Welsh landscapes since woodland clearance. In: Boardman, J., Foster, I.D.L. and Dearing, J.A. (eds.) *Soil Erosion on Agricultural Land*, John Wiley and Sons, Chichester.

Evans, R., 1993. Sensitivity of the British landscape to erosion. In: Thomas, D.S.G. and Allison, R.J. (eds.) *Landscape Sensitivity*, John Wiley and Sons Ltd, Chichester.

Evans, R., 1996. *Soil Erosion and its Impacts in England and Wales*, Friends of the Earth, London.

Evans, R., 1997. Soil erosion in the UK initiated by grazing animals, *Applied Geography*, **17**, 127-141.

Gilpin, L.R., 1997. *Grange-over-sands: a Resort and its Railway*, Cumbrian Railways Association, Grange-over-Sands.

Gooderson, P.J., 1970. Railway construction in mid-nineteenth century Lancashire: a study based on the diary of James Stelfox 1855-70, *Transactions of the Historical Society of Lancashire and Cheshire*, **121**, 137-151.

Gray, A.J., 1972. The ecology of Morecambe Bay v. the salt marshes of Morecambe Bay, *Journal of Applied Ecology*, **9**, 207-220.

Gray, A.J. and Adam, P., 1974. The reclamation history of Morecambe Bay, *Nature in Lancashire*, **4**, 13-20.

Gray, A.J. and Scott, R., 1987. Salt marshes. In: Robinson, N.A. and Pringle, A. (eds.) *Morecambe Bay: An Assessment of Present Ecological Knowledge*, Resource Paper for the Centre for North-West Regional Studies in Conjunction with the Morecambe Bay Study Group, University of Lancaster.

Harris, A., 1962. The seaside resorts of Westmorland and Lancashire north of the sands in the nineteenth century, *Transactions of the Historical Society of Lancashire and Cheshire*, **115**, 147-162.

Inglis, C.C. and Kestner, F.J.T., 1958. The long-term effects of training walls, reclamation and dredging on estuaries, *Proceedings of the Institute of Civil Engineers*, **9**, 193-216.

Jarvis, R.C., 1948. The appointment of ports in Cumberland, Westmorland and Lancashire-North-of-the-Sands, *Transactions of the Cumberland and Westmorland Antiquarian and Archaeological Society*, **47**, 128-165.

Jenkins, J., (ed.) 2002. *Remaking the Landscape: the Changing Face of Britain*, Profile Books, London.

Kestner, F.J.T., 1956. A study of the erosion and accretion during cyclic change in an estuary, *Journal of Agricultural Engineers Resources*, **1**, 63-67.

Kestner, F.J.T., 1970. Cyclic changes in Morecambe Bay, *Geographical Journal*, **136**, 85-97.

Kestner, F.J.T. and Inglis, C.C., 1956. A study of erosion and accretion during cyclic changes in an estuary, *Journal of Agricultural Engineers Resources*, **1**, 63-67.

Lucas, J., 1744. *A History of the Parish of Warton (Compiled 1710-1740)*, Lancaster. Reprinted and edited in 1931 by Ford, J.R. and Fuller-Maitland, J.A., 1931. Titus Wilson and Sons, Kendal.

Limbrick, K.J., Whitehead, P.G., Butterfield, D. and Reynard, N., 2000. Assessing the potential impacts of various climate change scenarios on the hydrological regime of the River Kennet at Theale, Berkshire, south-central England, UK: and application and evaluation of the new semi-distributed model, INCA, *The Science of the Total Environment*, **251/252**, 539-555.

Loaiciga, H.A., Valdes, J.B., Vogel, R., Garvey, J. and Schwarz, H., 1996. Global warming and the hydrological cycle, *Journal of Hydrology*, **174** (1/2), 83-127.

Marker, M.E., 1967. The Dee estuary: its progressive silting and salt marsh development, *Transactions of the Institute of British Geographers*, **41**, 65-71.
Meeus, J.H.A., Wijermans, M.P. and Vroom, M.J., 1990. Agricultural landscapes in Europe and their transformation, *Landscape and Urban Planning*, **18**, 289-352.
Pringle, A.W., 1995. Erosion of cyclic salt marsh in Morecambe Bay, North-West England, *Earth Surface Processes and Landforms*, **20**, 387-405.
Pringle, A.W., 2003. Recent rapid erosion of Silverdale salt marsh, *Keer to Kent*, **48**, 1-17.
Robson, A.J., Jones, T.K., Reed, D.W. and Bayliss, A.C., 1998. A study of national trend and variation in UK floods, *International Journal of Climatology*, **18**, 165-182.
Sansom, A., 1996. Floods and sheep – is there a link? *Newsletter of the British Hydrological Society*, **49**, 1-4.
Stockdale, J., 1872. *Annals of Cartmel*, Reprinted in 1978 by Michael Moon, Beckermet, Cumbria.
Stopford, S.C., 1951. An ecological survey of the Cheshire shore of the Dee estuary, *Journal of Animal Ecology*, **20**, 103-122.
Theaker, J.D., 1981. *Coastal Change and Man in the Grange-Silverdale Area of Morecambe Bay*, Unpublished dissertation, Department of Geography and Geology, Huddersfield Polytechnic.
Walvin, J., 1978. *Beside the Seaside. A Social History of the Popular Seaside Holiday*, Lane, London.
Watson, A., 1985. Soil erosion and vegetation damage near ski lifts at Cairngorm, Scotland, *Biological Conservation*, **33**, 363-381.
Waugh, E., 1861. *Rambles in the Lake District*, Whittaker, London.
Whitby, M. (ed.) 1992. *Land Use Change: the Causes and Consequences*, HMSO, London.
Williamson, T., 2000. Understanding enclosure, *Landscapes*, **1**, 56-79.
Wilson, P.N., 1973. Kendal reservoirs, *Transactions of the Cumberland and Westmorland Antiquarian and Archaeological Society*, **73**, 325-344.
Wilson, A.D. and Harrington, G.N., 1984. Grazing ecology and animal production. In: Harrington, G.N., Wilson, A.D., and Young, M.D. (eds.) *Management of Australia's Rangelands*, CSIRO, East Melbourne.
Yelling, J.A., 1977. *Common Field and Enclosure in England 1450-1850*, Macmillan Press Ltd, London.

Chapter 4

# Land Management, Cadastral Reform and Biodiversity: A New Zealand Study

Munir Morad

## Introduction

New Zealand's cadastral arrangements have come under scrutiny in recent years, prompted by concerns of over rapid loss of native biodiversity and urban expansion. A rise in the number of Maori land claims over the past two decades, and the enactment of the historic environmental legislation, the Resource Management Act of 1991, have particularly focused attention on the need for cadastral reform. However, land managers and legislators recognised from the outset that any reforms would have to reconcile seemingly incompatible goals, by being legally sound, culturally appropriate, and environmentally sustainable.

Most of the land in New Zealand is now subject to environmental legislation that incorporates concern for both environmental sustainability and Maori cultural values and ancestral rights. The presence of this legislation, and of a growing capacity among Maori to be involved, means that Maori concepts of land management and sustainability ought to receive more attention, as these are likely to influence the shape of future cadastral reforms for Maori lands. However, Maori lands tend to be the most obscure pieces in New Zealand's cadastral jigsaw, because most partitions and cadastral adjustments remain unsurveyed for years. Maori land issues therefore pose a considerable cadastral challenge. Numerous legislations and commissions of inquiry have been formed over the past decades to address the problem, but progress has been slow. Cultural factors have often been cited to explain the failure to resolve the persistent legal and technical hurdles confronting Maori land issues. As a New Zealand Judge has observed:

> the European land tenure system, in its present form, is a veritable engine of destruction of Maori land holdings (Winmill and Morton, 1993).

The problems facing Maori land tenure reform in New Zealand is highlighted by the fact that when the Treaty of Waitangi was signed (in 1840), Maori owned nearly 99 per cent of Aotearoa/New Zealand land. Today, the percentage is nearer the 5 per cent mark (less than half their population share).

Nearly 1.2 million hectares of Maori land (approximately 5 per cent of New Zealand land) was confiscated in the second half of the nineteenth century from tribes 'believed to have been in rebellion'. Much of this confiscated land has since found its way into private titles, which are not covered by modern compensation schemes (Asher and Naulls, 1987).

The complex cadastral situation of Maori land has evoked much discussion in legal, surveying, and environmental planning circles in New Zealand. Non-Maori experts have often argued that progress can be made through a programme of title improvement, whereby all of the unsurveyed Maori land that was partitioned is surveyed. However, for many cultural and economic reasons, Maori are reluctant to accept such arrangements. This is largely because Maoritanga (Maori culture) is steeped in communal values, and the prospect of freehold ownership is impractical where multiple ownership and successions would involve fragmenting a fast shrinking acreage held in small and isolated parcels (Morad and Jay, 1997).

The current cadastral complexity has arisen from a legacy marred by land confiscations, the intensification of agricultural production, and an apparent intolerance of Maori land tenure. A solution of these problems will have to embrace modern cadastral, environmental and economic standards, as well as acknowledge Maori cultural values. As a Maori scholar studying the history of Maori land tenure system once observed (Kawharu, 1977, pp. 310-311):

> Choices open to tribal groups and to the individual were much as they had ever been. But since the Treaty of Waitangi in 1840 they have become increasingly qualified by the operations of a sophisticated market economy. And whatever the action taken by the Maori in terms of these qualified choices, almost all have lent their own special impetus to the disintegration of his community life. Yet land still offers him opportunity, both for raising his material level of living and for maintaining his cultural identity as a Maori.

Resource managers are now required by legislation to consider the values and concerns of Maori in relation to land and Maori are developing an increasing capacity to be involved. This process is likely to have flow-on implications for land information system requirements as the values and concerns of Maori become recognised and recorded for land management purposes. According to the Resource Management Act (1991), environmental agencies are required to:

> recognise and provide for ... the relationship of Maori and their culture and traditions with their ancestral lands, water, sites, waahi tapu, (sacred sites) and other taonga (treasures).

## Land biodiversity issues

Long before New Zealanders heard of globalisation and thought of its ramifications, they had already globalised a large proportion of their country's ecosystems, transforming them into entities totally dependent on the import of nutrients. Apart from the obvious ecological risks involved in this trend, this level of dependency on nutrient imports suggests that continuing maintenance of the country's soils is at risk from potential disruption of fertiliser supplies prompted by economic, trade or military crises (Bennett et al., 1999).

About 60 per cent of the area of New Zealand is currently in pasture, and the introduced (largely northern hemisphere) plants that form the basis of these pastures have generally required more soil nutrients than the local soils have been able to provide. High lime and fertiliser rates have been needed to make these plants feel at home, and it has taken a massive effort by the farming industry to transform and maintain the fertility of the country's soils. This degree of intervention has resulted in the development of new soil ecosystems that form the basis of New Zealand's relative affluence. The new ecosystems are, however, strongly dependent on the continued importation of nutrients, predominantly phosphorous. As Hewitt (1999) noted:

> When we observe the boundary between native ecosystems and derived pastoral agroecosystems, we can readily appreciate the degree of changes that have occurred above ground, and we know about some of the more rapid changes that have occurred in the topsoils: changes to soil organic matter and nutrient levels. But we have no comprehensive understanding of the changes that have occurred below ground. What have we really done over the past 150 years to our soils? One thing we can be reasonably sure of, is that because many soil attributes adjust very slowly to changes above ground, it is likely that soils are still adjusting to changes in land cover that occurred last century and that consequential changes will continue.

In addition to the wholesale transformation of the physical landscape, the new arrivals to New Zealand's shores brought with them a host of intended and unintended visitors, including agricultural, horticultural and ornamental plants, farm livestock and domestic pets, but also rats and other pests. Rats, first introduced by Captain Cook in 1770, quickly spread to all corners of New Zealand, so that by 1870 visitors to Stewart Island reported them in abundance (King 1984, p. 69).

The impact of these immigrant plants and animals has been subtle but long lasting. They have brought about changes in the composition, structure, and function of indigenous vegetation and indigenous ecosystems, and the reduction of indigenous biodiversity (Morad and Jay, 2000). By the mid-20th century, their impact was more widespread. Rabbits in the South Island high country, together with sheep, had helped to create conditions that encouraged the spread of hieracium and other problem plants. The combined attack of goats, deer, and possums had impoverished large areas of indigenous forest, while ibex, and other introduced alpine mammals had diminished the native biodiversity of large areas of alpine herb fields. In the words of Crosby (1986, p. 268):

> the forces … reshaping New Zealand did not maintain their headlong pace and make a Europe out of New Zealand, but they did confirm it as a Neo-Europe.

This transformation of the colonial landscape, much of it unregulated, was often tolerated if not encouraged by government policies. According to Memon (1991, p. 26):

> massive transformation of the rural landscape, consequent upon the development of a monocultural system of pastoral and forestry land uses, land clearance and land drainage, often promoted by generous central government subsidies and grants, took place outside the scope of the town and country planning process.

Despite a widespread perception to the contrary, productive soils in New Zealand are rare, and account for as little as 5 per cent of the country's land resources (Hewitt, 1999). From an ecological point of view, it is important to be aware of this situation as identifying sustainable land productivity allows decision makers to evaluate the economic value of any land area, its versatility, production capacity and environmental (protection) value. Similarly, knowing the geographical distribution of good quality soils would militate against their neglect; and local authorities can develop Regional Policy Statements and District Plans with greater effectiveness. Traditionally, New Zealand land managers have been mainly concerned with soil quantity, focusing on erosion control. Now, however, regional environmental policies also require a focus on soil quality (so-called 'soil health'). Land managers must maintain the quality as well as the quantity of their soil resource, and there are national and international reasons why the maintenance of biodiversity is a desired outcome for New Zealand policy makers.

Some loss of soil biodiversity may be recoverable, but this assessment is subject to many complicating factors. Soil structures and organic matter levels tend to recover under rotations, but the loss of soil flora and fauna can take hundreds of years to recuperate, and require prolonged, careful management. Another complication that could impede soil recovery is the somewhat unpredictable pattern of soil response to contamination. The soil's capacity to filter contaminants depends on, among other things, acidity and organic matter levels. If contaminated soils become acidified, or if they lose organic matter, they may switch from being a contaminant sink to a contaminant source (Sparling et al., 1998). This process, which has been termed by some scientists as the 'chemical time bomb' effect, has received considerable attention in New Zealand in recent years.

Loss of soil biodiversity may be masked by management and not noticed by the farmer (Bennett et al., 1999). It is possible to continue arable cropping for long periods, with reasonable production levels, while running down soil health. Before yields are affected, input costs are likely to rise but profit levels may continue to be adequate. It is therefore relatively easy to recognise and mitigate the more obvious forms of soil erosion, while other degradation losses, notably those affecting soil biodiversity, may be obscure, gradual and in time irrecoverable. Soil biodiversity losses may be masked for a time by cultivation and fertiliser supply, but their build-up is subject to long lag times with unpredictable consequences. New Zealand soil scientists have made good progress in choosing soil health indicators. However, they must now make progress in understanding the dynamics of changes induced by loss of biodiversity, in order to predict the lag times involved between pressure and response. Furthermore, they must anticipate soil resilience to specific land management practices, and judge the significance of change in terms of effects on soil ecosystem functions (McIntrye et al., 1996).

Soil plant and animal species evolve and adapt to the physical conditions of their environment over long periods of time, and in turn modify their ecosystem. Ecosystems and communities are therefore characterised by dynamic biological and physical processes, and an understanding of these processes is important for managing biodiversity. Healthy resilient ecosystems are important for maintaining the resources we need (clean air, water and soils), mitigating the affects of pollution, and resist pests and diseases. Healthy soil ecosystems are equally instrumental for economic, social, cultural and ecological well being. New Zealanders cannot afford to be complacent about their natural heritage, including their country's soil biodiversity, as large tracts of the countryside have become devoid of their indigenous icons. Like most industrialised states, New Zealanders have tended to subordinate nature – the islands' indigenous environments – to production and economic prosperity.

## Community issues and legislation

The conservation of biological diversity is increasingly recognised by environmental planners as an essential component of sustainable regional development. Australasian ecologists are generally well qualified to contribute to the development and implementation of biodiversity conservation plans and strategies. As signatories to the 1992 Earth Summit agreement, Australia and New Zealand have been obliged to prepare "national strategies, plans or programmes for the conservation and sustainable use of biological diversity" (Convention on Biodiversity, 1992, Article 6).

Local government is important for biodiversity conservation in a number of ways:

- by providing a legal mandate to promote environmental protection at local and regional level;
- it is accountable to individuals and communities for environmental conditions within their local area;
- it can harness community involvement in environmental action; and
- it is potentially the level of government that can provide the ongoing care that is necessary for long-term ecological protection and restoration.

In New Zealand, district plans produced by territorial local authorities have thus been obliged to make provision for the protection of native habitat. Techniques include the use of schedules of ecologically significant sites, restrictions on the clearing of native forest, and provisions for encouraging the protection or restoration of riparian margins. Experience has shown that where the skills of planners in relation to community consultation have been fully involved, community acceptance of provisions for habitat protection has been much stronger than in situations where local or regional government have imposed such provisions without community consultation.

New Zealand has incorporated the principle of biological diversity within the government's Environment 2010 Strategy. The Strategy includes, as one of its main aims, the protection of:

> indigenous habitats and biological resources by: maintaining and enhancing the net area of New Zealand's remaining indigenous forests and enhancing the ecological integrity of other remaining indigenous ecosystems; promoting the conservation and sustainable management of biological diversity so that the quality of our indigenous and productive ecosystems is maintained or enhanced.

The legislative power of local government in New Zealand is provided through a number of statutes including the Resource Management Act 1991 and the Local Government Act 1974. These laws encourage a degree of environmental responsibility, which is particularly important for biodiversity protection because it extends to land in private ownership. Given that most of the conservation land in public ownership is generally over 300 metres above sea level, such areas tend to be representative of higher altitude ecosystems, as lower altitude ecosystems are often in private ownership. The Resource Management Act 1991 applies to all privately owned land and controls the development and use of air, water, soil, land and associated natural and physical resources (apart from minerals), including native plants, animals and ecosystems. The Act states, as a matter of national importance, that:

> all persons exercising functions and powers under it ... shall recognise and provide for ... the protection of areas of significant indigenous vegetation and significant habitats of indigenous fauna.

The Treaty of Waitangi, the Resource Management Act 1991, the Historic Places Act 1993, and the Maori Land Act 1993 require consideration of Maori cultural, historic, spiritual and physical values in environmental and social planning. Such values include whi tapu (sacred sites), marae (Maori community halls), and natural resources such as geothermal areas and indigenous flora and fauna. At present these legislative requirements cannot be met fully because suitable information is not available. Resource managers are now required by law to consider the cultural values and concerns of Maori in relation to land and Maori are developing an increasing capacity to be involved. This process is likely to increase as the legislation becomes entrenched, and will have flow-on implications for environmental planning requirements as the concerns of Maori become recognised in devising regional plans for biodiversity conservation and landscape protection.

However, experience suggests that legal instruments alone are seldom sufficient to encourage greater environmental responsibility, especially as most such legislation tends to be 'effects-based' (Gibellini 2001, p. 13). Of equal relevance is providing incentives for conservation. Froude (1997, pp. 17-20) observed that landowners tend to react negatively to regulatory mechanisms of conservation, and prefer positive approaches such as incentives and provision of information. Landcare groups have been initiated by some councils as a means of promoting environmental education and individual motivation, and have been shown to be effective (Environment Waikato, 1997, p. 40).

## Conclusion

This chapter argues that, in the light of growing acceptance of the concept of sustainable development, land managers can and should adopt methods and principles of planning and design that protect landscape heritage and native biodiversity. The loss of natural landscape habitats has become a matter of increasing concern at global, regional and local levels. It is a particular problem within New Zealand because of the high rates of endemism characteristic of New Zealand species, and their vulnerability to habitat loss and the effects of introduced competitors. Landscape conservation has become widely accepted as a key element of environmentally sustainable development.

The government of New Zealand was a major player in the advent of the UN Convention on Biological Diversity, and has pledged a commitment to promote biodiversity conservation through the protection of natural landscape habitats. Local and regional government can also be helpful in this regard by bringing about bylaws to mitigate the loss of native landscape habitats, as they are the levels of government that most directly affect actions of private landowners and managers on the ground, and are most directly accountable to local communities. The Local Government Act empowers local and regional government to raise funds to purchase land for purposes such as riparian protection, while the Resource Management Act allows districts to impose regulations and conditions in certain circumstances, for example, at the time of land subdivision.

In a world where environmental conflicts and economic pressures are likely to grow, planning for preserving landscape heritage requires new knowledge and skills in relation to ecosystem processes and species biology. In New Zealand, the requirement for greater understanding of biological systems must be matched by a commitment to involve indigenous Maori in the planning process. Future protection of native species and ecosystems is likely to involve the development of systems of co-management where central government (in the form of the Department of Conservation) and local authorities are prepared to trust local Maori land-owning groups with the management of local biological resources.

Land use planners and soil managers assist in moving concepts and principles to policies and practice, by working to integrate the human considerations with the ecological. They can determine which areas of land use conflict are based on substantive differences of interest, and which are based on ignorance or lack of information about the ecological value of a landscape. In cases of major conflict, the resolution of differences can often be achieved by analysing elements of the landscape in terms of their relative value for the wider community. Where it turns out that some fragile environments are particularly important for the conservation of native habitats, for example, land use planners and soil managers can include provisions within plans of action that will seek to bring about the protection of such areas, by regulation or incentive.

However, while the relationship between land and people is relatively dynamic in most societies, it has not been always possible in New Zealand to treat land as a well-defined economic asset subject to market mechanisms. In both indigenous and Western cultures, land is a primary element of the social and ecological fabric, and a crucial link in a delicate balance between economic, environmental and cultural agencies. The lack of recognition of cultural values and other institutional conditions is probably the most common cause for the intricate functioning of New Zealand's cadastral system. The land information system in New Zealand has been developed by the politically dominant settler culture as a structured arrangement, which places great emphasis on private property rights. The clear and exact identification of properties has gone hand in hand with the development of a property records system that allows trading of land parcels as commodity units. A land information management (LIM) system based on tenure fitted well with the economically-active migrant culture (in which historical roots are relatively fresh); but difficulties have inevitably arisen in trying to apply this system to fuzzier notions of ownership and boundary in many parts of New Zealand's 'open' lands as well as territories traditionally claimed by Maori communities.

The reform of New Zealand's cadastral system can, therefore, only be achieved by aiming to serve the different client groups that are affected by the gathering or supply of land information. From a management perspective, it is very important to appreciate the legal, cultural and environmental constraints under which the prospective land information system will work, in order to obtain the support of the client community. It is scarcely useful to draw a land information management strategy that is not backed up by legal and cultural considerations.

## References


Asher, G. and Naulls, D., 1987. *Maori Land*, New Zealand Planning Council, Wellington.

Bennett, R., Meister, A.D. and Wilkinson, R.L., 1999. Sustainable soil management in New Zealand: farmer beliefs, attitudes, and motivations, *Discussion Paper in Natural Resource and Environmental Economics, No. 21*. Centre for Applied Economics and Policy Studies, Massy University, New Zealand.

Crosby, A.W., 1986. *Ecological Imperialism, The Biological Expansion of Europe, 900-1900*, Cambridge University Press, Cambridge.

Environment Waikato, 1997. *Regional Policy Statement*, Environment Waikato Regional Council. Hamilton, New Zealand.

Froude, V., 1997. Biodiversity protection provisions, *Planning Quarterly*, **126**, 17-20.

Gibellini, L., 2001. The challenge of sustainability, *Survey Quarterly*, **27**, 13-17.

Hewitt, A.E., 1999. Keynote Address: Our valuable soils, *Manaaki Whenua Conference*, Te Papa, Wellington, New Zealand, April 21 1999.

Kawharu, I.H., 1977. *Maori Land Tenure*, Oxford University Press, Oxford.

King, C., 1984. *Immigrant Killers: Introduced Predators and the Conservation of Birds in New Zealand*, Oxford University Press, Oxford.

Memon, A., 1991. Shaking off a colonial legacy? Town and country planning in New Zealand 1870 to 1980s, *Planning Perspectives*, **6**, 19-32.

Morad, M. and Jay, M., 1997. Cadastral reform of indigenous land information, and environmental sustainability in New Zealand, *Surveying and Land Information Systems*, **57**(2), 120-123.

Morad, M. and Jay, M., 2000. Protecting New Zealand's native biodiversity, *Biologist*, **47**(4), 1987-201.

McIntrye, S., Barrett, G.W. and Ford, H.A., 1996. Communities and ecosystems. In: Spellerberg, I.F. (ed.), *Conservation Biology*, Longman, London.

Sparling, G., Schipper, L. and Hewitt, A., 1998. Soil quality characteristics of Waiareka and Wakanui soils under cropping, *New Zealand Soil Science Society Conference Abstracts*, Gisborne, 16-19 November, 153-154.

Winmill, R.H. and Morton, R., 1993. The implication of cadastral reform for Maori land, *New Zealand Surveyor*, **283**, 28-39.

# Chapter 5

# The Contested Territories of Costa Rica: Sustainable Land Management Challenges in the Neotropics

Simon Evans

## Introduction

This chapter investigates the complex processes associated with deforestation in the neotropics and discusses the challenges inherent in addressing forest loss within a sustainable development context. Utilising Costa Rica as a major case study example, the chapter identifies a range of disparate pressures exerted upon forest resources in the Central American region and considers the efficacy of remedial methods employed to counter their destructive impacts. This task requires decision makers and land managers to adapt conventional land use structures to reflect sustainable motivations and local realities, allied to a strong political commitment on the part of both the international community and national governments to ensure that current pressures are not compounded by the lure of short-term economic advantage. The need to evaluate the validity of existing land management structures represents a key starting point for this process, one which may ultimately necessitate radical and innovative models of land reform to ensure more equitable access to the natural resources traditionally utilised by indigenous and local communities. In order to challenge the hegemony imposed by institutional decision-making hierarchies, it is vital that both current and past land use structures are reassessed to determine their contribution to the achievement of preferred sustainable development outcomes. Whilst many traditional land management structures have all too often been dismissed as primitive and backward (Robinson and Bennett, 2000), the challenge posed by sustainability involves the reintroduction of decentralised structures in which the active participation of local people is not only legitimised but is actively sought and encouraged. This requirement enhances the influence of non-governmental organisations and reflects a visible shift in policy orientation from economic exploitation and production to conservation and environmental regulation.

## Setting the scene

The emergence of sustainability as a dominant development paradigm has increasingly brought land management systems under scrutiny. The emphasis upon management to achieve *either* production *or* protection of natural resources has created numerous inequalities in environmental quality and performance. This has been evident from a variety of economic, socio-cultural, physical, and political perspectives. This scenario reflects the extent of competing interests involved in the utilisation of land resources and the concomitant complexities generated by the diverse demands and values attached to the multiple usage of land. Where land is limited, these demands increasingly conflict, with economic criteria frequently employed to ascertain the relative merits of different land management solutions. In order to maximise the productive potential of such land, intensive management structures are commonly introduced to ensure optimum economic efficiency and maximised profits. In many forest environments, the returns on investment provided by non-timber activities are frequently perceived to exceed the existence value of the forest itself, thus prompting its clearance for alternative uses.

As forest resources become exhausted, the values attached to them assume incremental significance the scarcer they become, particularly when viewed from the safe distance of a dominant western environmental paradigm (Weaver and Oppermann, 2000). Equally, at the local level the ability of forest dependent communities to satisfy their basic needs becomes compromised as their access to traditional products and services is eroded. A logical progression of this debate culminates in concerted demands to halt forest loss in order to secure the conservation of natural resources. This is primarily pursued through the conveyance of protected area status upon sites of particular aesthetic and ecological significance, aided by the infusion of international funding. While this approach is habitually equated with the achievement of sustainable development goals, evidence would suggest that protected area designation might in certain instances, constitutes an actual cause of deforestation rather than a panacea (Brandon et al., 1998). The application of protected area status has, however, been enthusiastically pursued in many developing nations, often to the detriment of local needs and demands. This is primarily due to the adoption of a generic model of protection, formulated in the developed world and applied universally with little consideration of social, cultural or political distinctiveness. This model is based on a preservationist ethic that nature and people should be spatially separated to ensure the continuation of natural processes beyond the influence of human intervention (Colchester, 1994; Shaw-Bond, 2000). Pursuance of this ideal requires the forced displacement of forest dwellers and the removal of traditional lifestyles, particularly those based around hunter-gatherer societies (Evans, 2000). The ensuing conflict that develops between preservation and utilitarian approaches to natural resource protection lies at the heart of deforestation debates concerning Costa Rica, an issue that needs to be considered alongside the broader issues associated with global responses to locally experienced problems. This chapter seeks to challenge a number of popular misconceptions prevalent in such discussions, utilising a sustainability approach in organising and reconciling the myriad arguments applied to natural resource decline.

## The challenge of sustainability

The most widely cited definition of sustainability is that employed by the Brundtland Commission to describe:

> development that meets the needs of present generations without compromising the ability of future generations to meet their own needs (WCED, 1987).

Whilst the inception of such a laudable concept has been enthusiastically supported by a broad cross-section of global society, its translation into an unambiguous development framework has proved increasingly elusive in practice (Redclift, 1987). Much of this problem has been attributed to vagueness of expression and terminology, the underpinnings of sustainability contained within a series of long-winded, often oracular official documents (WCED, 1987; UNCED, 1992). This ambiguity has undoubtedly contributed to the appropriation of the term sustainability to justify developments and activities that would otherwise be considered inappropriate (Lele, 1991). The polarisation of arguments created by such uncertainty reflect a range of conflicting interpretations (Bridger and Luloff, 1999) impacts attributed which have proved difficult to counter despite the theoretical targeting of sustainability towards integration and holism. Whilst these shortcomings may compromise the effectiveness of sustainable development as an operating procedure, the notion of sustainability nevertheless offers a theoretical framework in which to organise the full inventory of inputs and to development activity, recognising their various interactions and interdependence. It is therefore important to identify the relationships and differences that exist between sustainability and earlier development philosophies.

The rationale behind sustainability differs significantly from preceding paradigms, demanding a holistic approach in determining the appropriateness of mechanisms designed to effect beneficial change. This commitment demands a universal view of processes and their manifold impacts, a consideration of systems as opposed to isolated assessments of individual factors, and their translation back to the local arena. The polarisation of debates within distinct disciplines has been a feature of many prior programmes, with policy makers finding it advantageous to limit the scope of potential outcomes within a rigid set of predetermined parameters. This approach is indicative of a reductionist world-view of development, one that reflects the application of positivist science. Informed by the Cartesian paradigm, these processes involve:

> breaking down components of a complex world into discrete parts, analysing them and then making predictions based on interpretations of these parts (Pretty, 1995).

Whilst this approach has tended to favour universal application over individual context, the competing notion of sustainability relies heavily upon holism and advocates the tailoring of appropriate developments to reflect local circumstances and imperatives, hence the slogan 'think global, act local'. In contrast to reductionist thought, holism is based upon the thesis that certain wholes represent more than the combined sum of their parts and that in order to judge the validity of different alternatives, an integrated systems approach is essential. In this case, a system can be viewed as a group of interrelated, interdependent, and interacting elements that together form a single functioning structure (Weaver and Oppermann, 2000).

When applied to the task of responding to environmental decline, this approach necessitates the simultaneous consideration of a range of economic, socio-cultural, physical, and political factors that jointly contribute to negative change. The generated information can then be utilised in the design of germane programmes that challenge the causes rather than manifestations of environmental deterioration. In this way, sustainable development can be viewed as a proactive measure that seeks to influence the design and implementation of appropriate models to minimise adverse effects throughout the extended life cycle of any given development or activity. The identification of a comprehensive inventory of stakeholder motivations represents an initial phase of this process. The differences between reductionist and sustainable archetypes can be illustrated through the consideration of agricultural systems based on the 'green revolution' model (Shiva, 1991) and the integrated 'agroforestry' approach (e.g. Rambo, 1984) considered in later sections of this chapter.

Returning to the issue of equity contained in the Brundtland definition, the central focus upon inter-generational equity should not disguise the importance of attaining intra-generational equity as a prerequisite of sustainable development planning. Merely identifying all relevant stakeholders is not sufficient and should represent just one facet of an approach that ultimately seeks to provide structures in which the contrasting motivations of disparate interests can be allocated equal access to decision-making structures. The notion of representative participation is therefore integral to this process. The following sections introduce the main case study of Costa Rica and seek to apply sustainability criteria in determining the causes of deforestation and the challenges inherent in their amelioration.

## Introducing Costa Rica

Located on the Central American isthmus, bordered by Nicaragua to the north and Panama to the south, Costa Rica covers a territory of 19,694 square miles, accommodating an estimated population of 3,340,000 people (Daling, 1998). Discovered by Columbus in 1502 during the course of his fourth and final voyage to the 'new world' (U.S. Department of State, 1998), Costa Rica's early development contrasted with that of its immediate neighbours due to the decimation of its indigenous population through a combination of war and disease. This situation frustrated colonial attempts to impose the *latifundia* system to the same extent as elsewhere in the region. This system consisted of the concentration of land in the power of a few influential owners who exploited the resources through the forced labour of a large dependent indigenous workforce (Place, 1993; Barraclough and Scott, 1987) and the concurrent expulsion of traditional owners from their land (Utting, 1993). The agrarian structure of Costa Rica evolved more equitably with a predominance of small and medium sized land holdings, although the later concentration of land by commercial enterprises provided a parody of this earlier approach (Weeks, 1985), an issue which is considered in later sections of this chapter. One consequence of this historical legacy was the retention of extensive tracts of largely undisturbed primary rainforest which, allied to a stable political structure, enabled Costa Rica to successfully market itself as a premier international ecotourism destination from the 1980s onwards (Fennell, 1999; Honey 1999).

It is impracticable to consider the political legacy of Central America and the attendant effects upon its disparate environments without reference to the proliferation of wars and territorial disputes that have broken out with alarming regularity in the region (Weinberg, 1991). Costa Rica has largely avoided these regional conflicts and, apart from a brief interregnum in 1948 when the Figueres regime staged a bloodless coup (Lara, 1995), stability has remained a feature of Costa Rican politics. Indeed, the Figueres government introduced a variety of policies that continue to influence Costa Rican life today, not least the wide-ranging social reforms and the abolition of the armed services (Edelman and Kenen, 1989). These changes contributed to the relative prosperity that Costa Rica currently enjoys and the emergence of a significant middle class, stimulating domestic tourism demands and exerting pressure upon the state to protect the nation's environmental heritage. On the other hand, critics point to the widening influence of the US (Karliner, 1990) as a consequence of such policies, militarily intervening in regional disputes (Honey, 1994) and treating the zone as little more than a refuse site (Hall et al., 1987) to support increasing levels of material consumption.

Ecologically, Costa Rica is characterised by its diversity and can be divided into six distinct regions (Moore, 1996), each containing a range of individual biomes spanning paramo, cloud forest, rainforest, dry forest, mangrove and coral reef (Blake and Becher, 1999). This ecological variance reflects Costa Rica's geographical positioning as a land bridge connecting North and South America (Janzen, D, 1983), containing vegetation and wildlife resources endemic to both continents. It also acts as a magnet to increasing numbers of ecotourists (INCAE, 1998, Lumsdon and Swift, 1998), the majority of which are accommodated within Protected Areas, which account for over one quarter of the national territory (Boza and Mendoza, 1981). Celebrated for its contribution to the international conservation agenda, Costa Rica is nevertheless subjected to the highest levels of deforestation anywhere in the western hemisphere (Repetto and Gillis, 1988). This paradox reflects the operation of a two-tier development approach adopted by successive governments to secure foreign revenue, namely production for export, and conservation to attract international funding and to facilitate subsequent ecotourism activity. The physical separation of these twin approaches has been mirrored in policy orientation, with procedures displaying an evident lack of integration and a subsequent inability to contain the effects of damaging land management processes.

## Demystifying deforestation

Contrary to many popularly held misconceptions, the causes of forest loss and decline cannot be attributed to any single influence or practice (Shiva, 1987; World Rainforest Movement, 1990), but to a plurality of complex pressures and demands (Barraclough and Ghimire, 2000). While the presence of logging companies may pose significant threats to parts of Asia, for example (Bevis, 1995), the timber industry exerts relatively little pressure on resources in Central America. Indeed, the majority of timber cleared in Costa Rica for the conversion of forest to cattle pasture was either burned off or left to rot, representing an overall loss of several billion dollars (Pedroni and Flores-Rodas, 1992 cited in Kuchli, 1997). This dissipative action takes on even greater significance when considered in the context of evidence suggesting that the withdrawal of a major US 'fast food' chain from Costa Rica in the 1980s resulted in state-owned banks writing off over $100 million in credits (Kuchli, 1997). Despite this, the continued use of subsidised credits as an incentive for promoting beef-related investment remains widespread in the region (Hecht, 1995). The following paragraphs aim to provide a comprehensive assessment of the manifold causes attributed to neotropical forest loss, documenting the interconnectedness of a range of multi-sectoral inputs and impacts.

Whilst some commentators apply Malthusian logic (e.g. Clark and Clark, 1995) to explain the processes of environmental decline, evidence from Costa Rica suggests a comparatively low rate of population growth (Daling, 1998) and thus an incomplete diagnosis. Indeed, a major indicator of population pressure relates to the expansion of agricultural zones to provide sufficient food to support a burgeoning population. Whilst the agricultural frontier continues to advance in Costa Rica, its primary motivation reflects the augmentation of the export sector rather than the achievement of locally utilisable crops. This approach is partially imposed upon Costa Rica as a consequence of international debt repayments and economic restructuring programmes prescribed by the major lending agencies (World Bank, 2000). In this way, the agricultural structure can be segmented into two discrete systems, namely the intensification of commercial enterprises for foreign exchange generation and the small scale activities associated with traditional farming methods for local consumption (Vandermeer and Perfecto, 1995). Commercial agriculture in Costa Rica can be subdivided further to reflect the primary influence of the banana, coffee and cattle raising sectors, each creating distinct physical territories within deforested zones. In each case, significant environmental damage can be attributed to the functions of such modernised agricultural systems. The following sections briefly introduce these key sectors, discussing their gradual accretion and combined impacts.

## Exogenous influences – the role of commercial agriculture

Dominated by the coffee, banana, and beef sectors, the conversion of land to commercial agriculture represents a major agent of deforestation in Costa Rica. Initially encouraged in the nineteenth century via government sponsored incentives to energise small-scale enterprise, the coffee industry rapidly emerged as a profitable concern, attracting the investment of a new class of wealthy 'coffee barons' (Rice, 1996; Gudmundson, 1986). As a production activity, coffee initially had little influence upon forest loss due to its reliance upon shade conditions provided by existing tree cover. The subsequent inclination of 'instant coffee' demands provided the impetus for engineering new strains of sun tolerant coffee plants (Segura and Reynolds, 1993), revolutionising the industry and initiating a rapid intensification of production techniques. This transformation increased the scale of forest clearance in the uplands and relied heavily upon the introduction of chemical inputs to management regimes seeking to maximise potential yields. This approach was mirrored in the lowlands through the parallel expansion of the banana industry, primarily in proximity to the Caribbean coast (Hall, 1985).

The banana industry emerged as a by-product of railroad development, utilising deforested corridors adjacent to railway tracks to produce fruit primarily for export to US markets (Hirsch and Aguilar, 1996), making full use of the trains for transportation purposes. Whilst both sectors contributed heavily to environmental damage (Mora, 1998), the fruit companies began to manipulate local labour reserves as demands for bananas escalated. Attracting labour from across the region during periods of peak demand and laying off excess workers as market fluctuations dictated became a feature of the fruit companies operations, leaving wage dependent former employees little option but to relocate. This power enabled the industry to impose rigid restrictions on the rights of workers, curtailing union representation to dissuade dissension and negate the threat of strike action (Brandon et al., 1998). The resulting diaspora increased the incidence of adapted forms of nomadic agriculture, an issue considered below, exacerbating illegal land occupation and prompting concomitant increases in environmentally destructive practices associated with gold prospecting and charcoal burning (Kuchli, 1997). Recent contractions in fruit production have emerged as a consequence of the so called 'banana wars', a protracted dispute between the US and Europe resulting from the European Community's decision to limit the import of 'dollar' bananas in favour of those produced in their former colonies and dependencies (Greenfeld, 1991).

If the spatial separation between coffee and banana production implies the existence of vast areas of intact primary forest in the intermediate zone, the introduction of cattle grazing throughout Costa Rica has effectively bridged this divide. The ramifications of beef production in Latin America has generated considerable debate over many years (Parsons, 1988) the 'hamburger connection' significantly influencing land use structures as multinational investors continue the conversion of large swathes of forest into grazing land to supply the 'fast food' trade (Nations and Komer, 1987). This form of land use differs from the intensive activities associated with coffee and banana producers, relying instead upon extensive areas of range to support low concentrations of cattle. Whilst the issues surrounding this process are diverse and wide ranging, for the purposes of this chapter, two important aspects emerge. Firstly, the removal of key nutrients from deforested lands and the compaction of surfaces preventing the re-colonisation of endemic vegetation (Beletsky, 1998) produces poor quality soils, reducing the carrying capacity of the range. Secondly, the idiosyncratic nature of Costa Rican law in only recognising title to lands that are being actively used, tend to legitimise and actually promote deforestation as a prerequisite for proving ownership. Furthermore, without supportive evidence of continued use, the land may be subject to future expropriation (World Conservation Monitoring Centre, 1992). These two issues illustrate the advantage of cattle grazing over alternative uses, enabling beef producers to accumulate, and justify perpetual ownership of, large tracts of open land, with cattle representing fluid 'live boundary markers' (Kuchli, 1997), delineating tenure parameters. As the range deteriorates over time, extensions of land are sought, resulting in the further consolidation of resources.

## The rise of the 'neo-colonial' agenda

The examples provided above are indicative of an economic reorientation from a subsistence-based to an export-based economy, undertaken in the name of progress and supported by a series of international loans. The resulting debt and the employment of an increasing proportion of Costa Rica's foreign earnings to satisfy repayment schedules have increased the economic and political influence of exogenous interests. Foreign governments and multinational investors, in turn, have been able to manipulate and increasingly guide national development structures to reflect their own priorities. In one sense, the political influence exerted over Costa Rica by US interventions into regional disputes has enabled the US government to monitor and dictate the expediency of power structures across Central America (Weinberg, 1991; Honey, 1994). Furthermore, those influences serve to actively expedite those powers considered to be sympathetic to their own national interests. In another sense, the economic powers conveyed upon multinational companies have created a dependence of traditionally self-reliant communities upon externally imposed waged labour opportunities. The subsequent ability of these companies to prescribe working conditions, remuneration levels and, should economic realities or labour relations dictate, to close down operations and relocate elsewhere has created a characteristically dependent and subservient workforce. As labour intensive operations are perpetually substituted with technological alternatives and market fluctuations influence production levels, labour requirements oscillate accordingly, with redundant workers forced to migrate to areas that have yet to be colonised or claimed.

As increasing amounts of land are granted official title, the abjuration of access to natural resources becomes an acute problem, creating pronounced inequalities between rival societal groupings, emphasising the widening gulf between the rich and poor. As the space available to dispossessed rural workers contracts so instances of illegal land occupation increase, magnifying the potential for conflict. In order to satisfy essential needs, itinerant landless classes are forced into a cycle of illicit land usage. Nomadic practices of this kind increase the potential for environmental deterioration as land management techniques that seek to parody traditional practices become intensified and temporally curtailed. This enforced dynamism compromises the customary balance sought between the twin objectives of extraction and conservation, expediting the exhaustion of primary resources, the destruction of valuable habitats and the disfigurement of cultural landscapes. This sets in motion a prolonged spiral of decline that further compromises the ability of the disenfranchised to maintain vital life support systems. An example of this form of distorted practice can be illustrated through the example of shifting cultivation.

## Indigenous influences

The practice of shifting cultivation (Colchester, 1995) represents a long established agricultural technique employed in many forest frontier territories. The periodic clearance of land for cultivation purposes is synonymous with traditional indigenous management regimes designed to ensure the rotational utilisation of renewable resources. The ability of forests to naturally regenerate following exploitation is well chronicled (e.g. Redford and Padoch, 1992) and is axiomatic of the argument that it is not the initial incursion into primary forest that dictates its irreversible decline but the subsequent deployment of the resulting land base. Based upon inherited knowledge and experiential practice, shifting cultivators have rotated the same series of land parcels for generations, leaving a requisite duration for the regeneration of resources before the cycle is repeated. This form of husbandry reflects the balance between conservation and exploitation, which can be translated into contemporary sustainable development debate as the prudent use of renewable resources. This process ensures that land fertility is controlled and production systems regulated, whilst the cut timber provides construction material, fuel and firewood, contrary to the perception that shifting agriculture relies exclusively upon wasteful 'slash and burn' operations (Colchester, 1995). Indeed, evidence suggesting that the majority of forest loss in Costa Rica has occurred since the 1960s (Dobson, 1996) acts to refute claims that small-scale farming operations represent the primary agents of deforestation (Lutz and Caldecott, 1996). In reality, this form of development imitates natural processes, undertaken by forest dependent communities aware of the need to avoid over-exploitation.

As argued previously, however, the consolidation of land under industrial ownership militates against such sustainable activity, reducing the time lapse employed between clearance cycles. This contributes further to the deterioration of resources and, as 'unmanaged' land becomes increasingly scarce, so the landless are pushed further into the remaining forest. The transformation of forested land from large, remote blocks to small, fragmented pockets creates a 'domino effect' as a result of the distension of the forest edge and the creation of an island biodiversity (MacArthur and Wilson, 1967; Diamond, 1975), attracting hunter-gatherer communities intent on allaying their subsistence deficits. Described as the 'edge effect' (Horta, 2000), this implies that the area under pressure from exploitation increases in proportion to the boundaries of previous clearances, including roads and access routes traversing otherwise intact forest zones. Disinherited peasants can view this as part of a wider spiral of decline, with the main incursions attributed to industrial interests, followed by uncontrolled forays. In an attempt to arrest such degeneration, Costa Rica embarked upon an extensive programme of natural resource protection. This involved conveying varying levels of protection upon 25 per cent of its land area, half of which is devoted to a network of 38 National Parks, Reserves and Wildlife Refuges (Blake and Becher, 1999).

## Protected area designation: preservation or destruction?

The Protected Areas programme adopted by the Costa Rican government has been championed by conservation organisations and government agencies alike as an exemplar of prudent environmental husbandry. Employing a series of classifications established within IUCN guidelines (Ceballos-Lascurain, 1996), core National Park areas are surrounded by a series of buffer zones which are subject to less stringent protective policy. Considered the antithesis of the production ethic, this preservation model has traditionally generated little resistance or critical debate due to its pervasive acceptance as an example of altruistic endeavour. Recent studies of the Corcovado (Brandon et al., 1998; Minca and Linda, 2000) and Tortuguera (Place, 1998) National Parks have, however, remedied this weakness to a certain extent, articulating threats to the inviolability of the parks system as external pressures become more acute. If the parks are indeed in peril, it follows that the profitable tourism market is also threatened.

Much of this problem can be attributed to the fact that the Costa Rican Protected Areas model is based upon a generic archetype originating from the pioneer parks initially established in nineteenth-century USA (West-Sellars, 1997), entailing as it does the forced resettlement of communities from traditional homelands. This is unsurprising considering the ubiquitous influence of North American scientists operating in Costa Rica at the time of their establishment (Bruggemann, 1997). This approach has commanded appreciable funding from the international community and in recent decades has benefited from the lucrative ecotourism market, becoming the nation's largest generator of foreign revenue (Weaver, 1998). The isolation of ecotourists within park boundaries, however, offers little opportunity for local inhabitants to financially benefit from foreign visitation and their forced exclusion from customary resources increases the prospect of future encroachment back into protected territories. The containment of this access is compromised by the imposition of Structural Adjustment Programmes initiated by international banking agencies to streamline debt repayments (Baker, 2000), necessitating a reduction in governmental budgets and staffing levels. This polemic reflects the polarisation of production and protection as independent sectors, the institution of protected area apparatus a reaction to rapid deforestation elsewhere in Costa Rica. When considering this procedure from a sustainability perspective, it could be argued that land use planning models continue to reflect a reductionist stance and that in order to address the protracted diminution of natural and cultural resources, a more holistic, integrated strategy is essential.

## The future?

The need for policy reconciliation between production and protection sectors represents a major challenge in Costa Rica. Formulation of a cohesive protocol recognising and reflecting the motivations and needs of all relevant stakeholders provides a key starting point for this process. This requires a greater autonomy for national government and a willingness on their part to transfer certain powers to local authorities who in turn should seek to empower the local populace to participate in the achievement of appropriate development and management structures. In order to accomplish this, a global commitment to dismantle inappropriate patterns of imposed development and a diminution of the power of multinationals in dictating the scope and nature of their activities is essential. In terms of forest resources it is crucial that decision makers examine the socio-cultural attributes of trees (e.g. Wolvekamp, 1999) in addition to their economic potential, recognising the diverse roles forests play in combating food insecurity, poverty and environmental degradation (FAO, 1990; Barraclough and Ghimire, 1995). This task requires the employment of robust accounting procedures capable of reconciling the multiple values attached to forestlands, mechanisms that exceed the conventional opportunity-cost arguments associated with foregone production. Equally important is the need to invest in rehabilitation and conservation efforts beyond protected area boundaries, seeking to enhance the environmental performance of whole regions rather than just specific sites (Evans, 2001). This should accompany changes in policy orientation to ensure that parks and people do not remain mutually exclusive entities, but instead provide for a broad range of subsistence demands to support local communities.

The example of Costa Rica illustrates a failure to fully realise the sustainable development paradigm shift in practice. A series of conflicting land uses, supported by abstracted policies, have led to the creation of a number of distinctly separate territories, clearly delineated from neighbouring zones economically, socio-culturally and physically. As available land becomes scarce and the livelihoods of increasing numbers of landless communities become further marginalised as a consequence, the natural resource base of the nation becomes seriously compromised. Ensuring a logical progression from policy formulation to integrated land use planning demands the design and implementation of innovative land management regimes that guarantee equitable access to local resources. At present, this sequence appears to be fractured and until this is remedied, it seems likely that the rapid pace of deforestation will persist and that social agitation will increase accordingly. The need to balance the multiple demands of society and natural resources, and of conservation and production, represents a priority in attempts to transform reductionist policies into sustainable outcomes. This task, it is argued, provides the basis for pursuing sustainable land management in the neotropics.

## References


Baker, K., 2000. Ecological possibilities and political constraints. In: Stott, P., and Sullivan, S. (eds.), *Political Ecology: Science, Myth and Power*, 157-178, Arnold, London.

Barraclough, S. and Ghimire, K., 1995. *Forests and Livelihoods: The Social Dynamics of Deforestation in Developing Countries*, Macmillan, London.

Barraclough, S. and Ghimire, K., 2000. *Agricultural Expansion and Tropical Deforestation: Poverty, International Trade and Land Use*, Earthscan, London.

Barraclough, S. and Scott, M., 1987. *The Rich Have Already Eaten: Roots of Catastrophe in Central America*, Transnational Institute, Amsterdam.

Beletsky, L., 1998. *Costa Rica: The Ecotravellers Wildlife Guide*, Academic Press, California.

Bevis, W., 1995. *Borneo Log: The Struggle for Sarawak's Forests*, University of Washington Press, Seattle.

Blake, B. and Becher, A., 1999. *The New Key to Costa Rica* (14th Ed.), Ulysses Press, Berkeley.

Boza, M. and Mendoza, R., 1981. *National Parks of Costa Rica*, INCAFO, Madrid.

Brandon, K., Redford, K. and Sanderson, S. 1998, (eds.), *Parks in Peril: People, Politics and Protected Areas*, Island Press, Washington D.C.

Bridger, J. and Luloff, A., 1999 Toward an interactional approach to sustainable community development, *Journal of Rural Studies*, **15,** 377-387.

Bruggemann, J., 1997. National parks and protected area management in Costa Rica and Germany: a comparative analysis. In: Ghimire, K. and Pimbert, M. (eds.), *Social Change and Conservation*, 71-96, Earthscan, London.

Ceballos-Lascurain, H., 1996. *Tourism, Ecotourism and Protected Areas*, IUCN, Gland.

Clark, D. and Clark, D., 1995. Life history diversity of canopy and emergent trees in a neotropical rainforest, *Ecological Monographs*, **62,** 315-344.

Colchester, M., 1994. *Salvaging Nature: Indigenous Peoples, Protected Areas and Biodiversity Conservation*, UNRISD Discussion Chapter No.55, United Nations Research Institute for Social Development, Geneva.

Colchester, M., 1995. Forest peoples and sustainability. In: Colchester, M. and Lohmann, L., *The Struggle for Land and the Fate of the Forests*, 61-95, Zed Books, London.

Daling, T., 1998. *Costa Rica in Focus*, Latin America Bureau, London.

Diamond, J., 1975. The island dilemma: lessons of modern biogeographic studies for the design of nature reserves, *Biological Conservation*, **7**, 129-146.

Dobson, A., 1996. *Conservation and Biodiversity*, Scientific American Library, New York.

Edelman, M. and Kenan, J. (eds.), 1989. *The Costa Rica Reader*, Grove Atlantic, New York.

Evans, S., 2000. Ecotourism in tropical rainforests: an environmental management option for threatened resources. In: Font, X. and Tribe, J. (eds.), *Forest Tourism and Recreation*, 127-142, CABI, Wallingford.

Evans, S., 2001. Community forestry: countering excess visitation in England's national parks. In: McCool, S. and Moisey, N. (eds.), *Tourism, Recreation and Sustainability*, CABI, Wallingford (In Press).

FAO, 1989. Forestry and food security, *Forestry Chapter 90*, Food and Agriculture Organisation of United Nations, Rome.

Fennell, D., 1999. *Ecotourism: An Introduction*, Routledge, London.

Greenfeld, K., 1999. Banana wars, *Business*, **153**, No.5, February 8.
Gudmundson, L., 1986. *Costa Rica before Coffee*, Louisiana State University Press, Baton Rouge.
Hall, C., 1985. *Costa Rica: A Geographical Interpretation in Historical Perspective*, Westview Press, Boulder.
Hall, B., Karliner, J. and Whitney, P., 1987. *Garbage Imperialism*, Earth Island, San Francisco.
Hecht, S., 1995. Brazil: landlessness, land speculation and pasture-led deforestation. In: Colchester, M. and Lohmann, L. (eds.), *The Struggle for Land and the Fate of the Forests*, 164-178, Zed Books, London.
Hirsch, J. and Aguilar, R., 1996. *Banana Expansion in the Humid Tropical Zone*, Lead international, www.lead.org/lead/training/international/costarica/96/ch5.html (Accessed 12 November, 2000).
Honey, M., 1994. *Hostile Acts: US Policy in Costa Rica in the 1980s*, University of Florida Press, Gainesville, Florida.
Honey, M., 1999. *Ecotourism and Sustainable Development: Who Owns Paradise?* Island Press, Washington D.C.
Horta, K., 2000. Rainforest: biodiversity conservation and the political economy of international financial institutions. In: Stott, P. and Sullivan, S. (eds.), *Political Ecology: Science, Myth and Power*, 179-202, Arnold, London.
INCAE, 1998. *Impacts on Developing Countries of Changing Production and Consumption Patterns in Developed Countries: The Case of Ecotourism in Costa Rica (Draft)*, Report to UNEP. INCAE, San Jose.
Janzen, D. (ed.), 1983. *Costa Rican Natural History*, University of Chicago Press, Chicago.
Karliner, J., 1990 Central America: political ecology and US foreign policy. In: Head, S. and Heinzman, R. (eds.), *Lessons of the Rainforest*, 144-155, Sierra Club Books, San Francisco.
Kuchli, C., 1997. *Forests of Hope: Stories of Regeneration*, Earthscan, London.
Lele, S., 1991. Sustainable development: a critical review, *World Development*, **19**(6), 607-621.
Lara, S., 1995. *Inside Costa Rica*, Interhemispheric Resource Centre, Albuquerque.
Lumsdon, L. and Swift, J., 1998. Ecotourism at a crossroads: the case of Costa Rica, *Journal of Sustainable Tourism*, **6**(2), 155-172.
Lutz, E. and Caldecott, J., 1996. *Decentralisation and Biodiversity Conservation*, World Bank, Washington D.C.
MacArthur, R. and Wilson, E., 1967. *The Theory of Island Biogeography*, Princeton University Press, Princeton.
Minca, C. and Linda, M., 2000. Ecotourism on the edge: the case of Corcovado National Park, Costa Rica. In: Font, X. and Tribe, J. (eds.), *Forest Tourism and Recreation*, 103-126, CABI, Wallingford.
Moore, D., 1996. *San Jose, Costa Rica: Urbanisation and Development*, Macalester College, www.macalstr.edu/~geograph/world-urbanization/dmoore/geography.html (Accessed 4 December 2000).
Mora, B., 1998. *Bananas for the World and the Consequences for Costa Rica* (mimeo), Foro Emaus, San Jose.
Nations, J. and Komer, D., 1987. Rainforests and the hamburger society, *The Ecologist*, **17**(4/5), 161-167.
Parsons, J., 1988. The scourge of cows, *Whole Earth Review*, Spring 1988, 40-47.
Place, S., 1993. *Tropical Rainforests: Latin American Nature and Society in Transition*, Scholarly Resources Inc. Wilmington.

Place, S., 1998. How sustainable is ecotourism in Costa Rica. In: Hall, C. and Lew, A. (eds.), *Sustainable Tourism: A Geographical Perspective*, 107-118, Longman, Harlow.
Pretty, J., 1995. *Regenerating Agriculture*, Earthscan, London.
Rambo, A,. 1984. Why shifting cultivators keep shifting: understanding farmer decision making in traditional agroforestry systems. In: UNDP, *Community Forestry: Some Aspects*, FAO, Bangkok.
Redclift, M., 1987. *Sustainable Development: Exploring the Contradictions*, Methuen, London.
Redford, K. and Padoch, C. (eds.), 1992. *Conservation of Neotropical Forests: Working from Traditional Resource Use*, Columbia University Press, New York.
Repetto, R. and Gillis, M., 1988. *Public Policies and the Misuse of Forest Resources*, Cambridge University Press, Cambridge.
Rice, R., 1996. Coffee, modernisation and ecological change in Northern Latin America, *Tea and Coffee Trade Journal*, 104-123.
Robinson, J. and Bennett, E. (eds.), 2000. *Hunting for Sustainability in Tropical Forests*, University of Columbia Press, New York.
Segura, R. and Reynolds, J., 1993. *Environmental Impact of Coffee Production and Processing in El Salvador and Costa Rica* (mimeo), UNCTAD/COM/20.
Shaw-Bond, M., 2000. Life in the balance, *Geographical Magazine*, **72**(9), 64-65
Shiva, V., 1987. *Forestry Crisis and Forestry Myths*, World Rainforest Movement, Penang.
Shiva, V., 1991. *The Violence of the Green Revolution*, Other India Press, Mapusa.
UNCED, 1992. *Resolution 44/228*, United Nations Conference on Environment and Development, New York.
US Department of State, 1998. *Costa Rica Background Notes*, www.state.gov/www/background_notes/costa_rica_0798_bgn.html (Accessed 21 November, 2000).
Utting, P., 1993. *Trees, People and Power*, Earthscan, London.
Vandermeer, J. and Perfecto, I., 1995. *Breakfast of Biodiversity*, Institute for Food and Development Policy, Oakland.
Weaver, D., 1998. *Ecotourism in the Less Developed World*, CABI, Wallingford.
Weaver, D. and Oppermann, M., 2000. *Tourism Management*, John Wiley, Brisbane.
Weeks, J., 1985. *The Economies of Central America*, Holmes and Meier, New York.
Weinberg, B., 1991. *War on the Land: Ecology and Politics in Central America*, Zed Books, London.
West-Sellars, R., 1997. *Preserving Nature in the National Parks*, Yale University Press, New Haven.
WCED, 1987. *Our Common Future*, Oxford University Press, New York.
Wolvekamp, P. (ed.), 1999. *Forests for the Future: Local Strategies for Forest Protection, Economic Welfare and Social Justice*, Zed Books, London.
World Bank, 2000. Costa Rica: forest strategy and the evolution of land use, *Evaluation Country Case Study Series*, World Bank, Washington D.C.
World Conservation Monitoring Centre, 1992. *Protected Areas of the World – A Review of National Systems: Costa Rica*, www.wcmc.org.uk/cgi-bin/pa (Accessed 12 November 2000).
World Rainforest Movement, 1990. *Rainforest Destruction: Causes, Effects and False Solutions*, WRM, Penang.

# Chapter 6

# Flood Defence and Estuary Management: a Case Study of the Blackwater Estuary, Essex, England

Peter Wynn

## Introduction

The Blackwater Estuary is an inlet of the North Sea, approximately 15 km long and 2.5 km wide at its mouth. On either side of the estuary, earth embankment seawalls protect a strip of reclaimed land of up to about 1 km in depth, normally constructed and maintained using locally dug clay. The history of seawalls in Essex is covered by Grieves (1959). Land use is mainly agriculture but the town of Maldon lies at the head of the estuary and Bradwell nuclear power station lies on its southern bank. The estuary is a major centre for water sport: there are sailing clubs on both sides of the estuary and marina facilities at Bradwell.

## Conservation significance of the estuary

The saltmarshes that fringe the estuary are an important habitat in their own right and, along with the mud flats, are internationally important for migratory bird species. Much of the estuary is now designated for protection either under various forms of international or national legislation, or is under the ownership of conservation bodies as indicated on Figure 6.1.

The Old Hall Marshes (627 ha) were designated as a Ramsar site under the International Convention on Wetlands. This designation has now been extended to include 4,395 ha of the marshes and mudflats in the estuary, which also form a Special Protection Area under the EU Directive on the Conservation of Wild Birds and a designated Site of Special Scientific Interest under the Wildlife and Countryside Act. The Essex Estuaries are also a candidate Special Area of Conservation under the EC Habitats Directive. The Royal Society for the Protection of Birds own Old Hall Marshes and this location, together with Tollesbury mud flats, form the Blackwater Estuary National Nature Reserve. The National Trust has property at Copt Hall and Ray Island, the latter being managed by the Essex Wildlife Trust who also lease Rat Island from Colchester Borough

Council. The Essex Wildlife Trust (2000) also report on their recent purchase of Abbotts Hall Farm on the north bank of Salcott Creek.

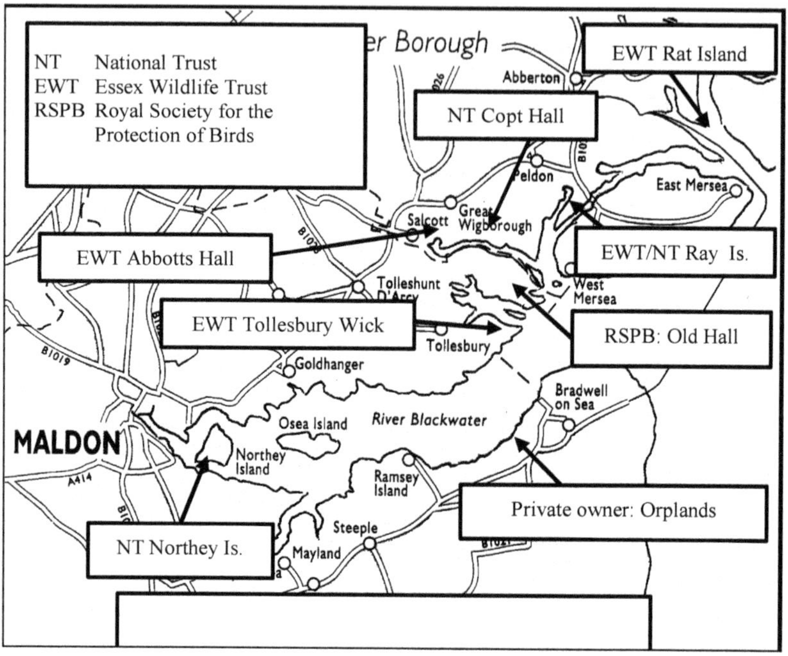


**Figure 6.1 Managed areas identified in text**

The estuary also forms part of the Essex Coast Environmentally Sensitive Area allowing farmers to receive payment for managing their activities in a manner that will protect areas of national landscape and wildlife importance. As will be discussed subsequently, these recent designations are significant in terms of the management of flood defence.

## Statutory basis of flood defence

Flood defence has a long legislative history but modern legislation concerning flood defence starts with the Land Drainage Act 1930, which gave powers to construct flood defence works to newly formed catchment boards. These were succeeded in turn by River Boards, River Authorities, Regional Water Authorities, the National Rivers Authority (NRA) and the Environment Agency. The flood defence functions of these bodies were carried out under the control of their land drainage (later flood defence) committees. There has traditionally been a strong agricultural representation on these committees and central government responsibility for flood defence also lies with the Minister of Agriculture, Fisheries and Food (MAFF).

Maritime coastal authorities, under the Coast Protection Act 1949, separately manage the protection of coasts along open sea against erosion. In 1985, administrative responsibility for this passed from the Secretary of State for the Environment to the Minister of Agriculture, Fisheries, and Food. Although most of the Blackwater Estuary lies in waters excluded from the provisions of the Coast Protection Act, this change in responsibility was part of a trend towards a unified approach to the coast, which has influenced management of the study area.

## Influences on flood defence 1950s to 1970s

In the 1960s, local authorities were encouraged by central government to prepare policies for the protection of coastal areas from over-development (Ministry of Housing and Local Government, 1963 and 1966). Essex County Council's (ECC) Interim Coastal Protection Policy (ECC, 1966) produced in response to these government requests covered an area up to five miles inland of the coast. The concept of Heritage Coast was introduced in the early 1970s (Department of the Environment, 1972) but no such designations were made in Essex. Traditionally, agricultural use of land has been outside of planning controls and these policies did not consider agricultural development.

Throughout the post-war period, the government through financial incentives encouraged the use of arable land. This resulted in the conversion to arable use of many of the pastures that fringed the estuary and led to a situation where higher seawalls could be economically justified. There was little requirement for conservation to be considered in undertaking flood defence works until the Water Act 1973 which required both the Minister of Agriculture, and Anglian Water Authority:

> in formulating or considering any proposals ... to have regard to the desirability ... of conserving flora, fauna ... of special interest ... and shall take into account any effect which the proposals would have on ... any such flora [or] fauna (Section 22[1]).

Figure 6.2 shows the inter-relationship of flood defence, agriculture, planning. and conservation that evolved under these circumstances.

## Developing concern over saltmarsh loss

During the 1970s, concern grew over the erosion of the saltmarsh fringe, which had previously reduced wave action and overtopping of the walls under storm conditions. This loss has resulted in increasing maintenance costs and it has become harder, on economic grounds, to justify capital schemes to improve seawalls protecting agricultural land. Dixon et al., (1998) reported a countywide loss of 2 per cent per year. From the late 1970s, this has led to trials of alternative 'soft engineering' techniques within and close to the Blackwater Estuary. At Deal

Hall, towards the south of the Dengie Peninsula, saltmarsh regeneration using the Schleswick-Holstein method (Kamps, 1962) of constructing polders with brushwood groynes and ditches was tested. A simplified trial using groynes alone was carried out at Cudmore Grove on Mersea Island, whilst at Marsh House, also on the Dengie Peninsula, an offshore wavebreak was constructed using disused lighters, and was followed by a similar scheme close to Sales Point at the entrance to the Blackwater. At Northey Island, on a site owned by the National Trust, the NRA carried out a scheme of managed retreat covering about 4 hectares. Leggett and Dixon (1994) reported that monitoring by English Nature showed that saltmarsh vegetation had re-established within two years of the work being carried out. The development of the inter-relationship between considerations of sea defence and conservation is shown in Figure 6.3.

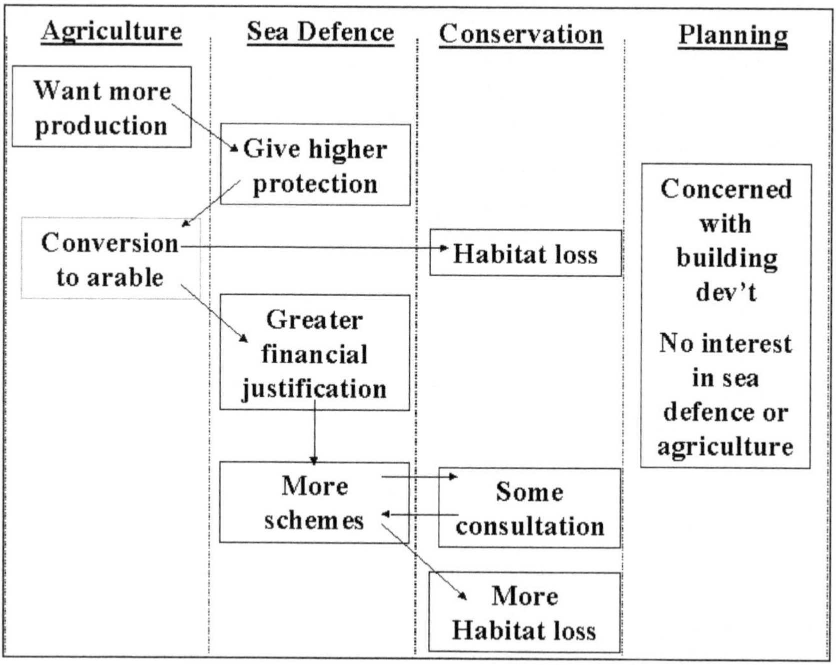


**Figure 6.2 Inter-relationship between sea defence, agriculture, conservation and planning up to 1970s**

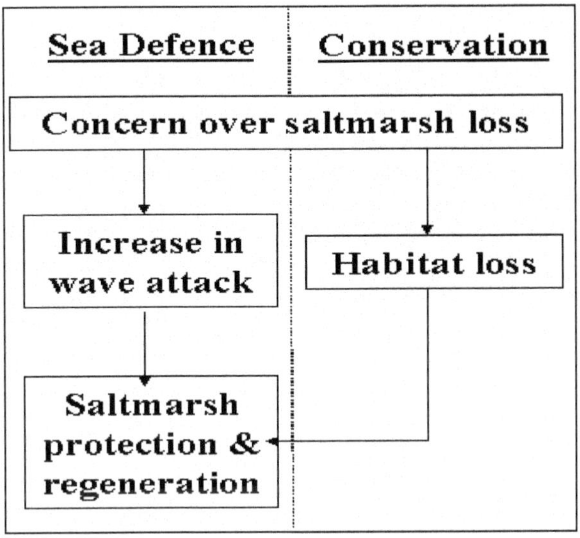


**Figure 6.3 Inter-relationship between sea defence and conservation in the late 1970s and early 1980s**

The Wildlife and Countryside Act 1981 also introduced a more active duty with respect to conservation on the Minister of Agriculture and Anglian Water, amending Section 22(1) of the Water Act 1973 by requiring them:

> in formulating or considering any proposals … so exercise their functions with respect to the proposals as … to further the conservation of flora, fauna…of special interest … and shall take into account any effect which the proposals would have on … any such flora [or] fauna.

Flood protection still did not feature strongly at this time in planning documents and ECC's interim coastal policy was subsequently replaced by Policy NR18 of the Essex Structure Plan (Essex County Council, 1982), which stated:

> There shall be the most stringent restrictions on development on the rural and undeveloped coastline outside built-up areas and any development which is exceptionally permitted shall not adversely affect the open and rural character or wildlife.

The boundary of the Coastal Protection Belt to which Policy NR18 applied was defined in a subsequent document (ECC, 1984). Account was taken of views of the coast and as far as possible the boundary of the coastal protection belt followed existing permanent features. It specifically stated that it did not deal with issues of sea defences or flood protection. No changes to Policy NR18 were made when the Structure Plan was subsequently reviewed (ECC, 1991).

## Developing national policies

By the early 1990s, the need for an integrated approach to coastal management, including defence from the sea and planning, had become generally accepted. This was partly the result of MAFF becoming responsible for both flood defence and coast protection and partly a response to potential sea level rise and the overall move towards sustainable development.

Moves were made towards a policy of management based on systematic study of lengths of coastline, which were later to become known as 'sediment cells'. The Blackwater Estuary formed part of a prototype major UK study of this kind (Fleming, 1989). The House of Commons Environment Select Committee also recommended an integrated approach to coastal management based on sediment cells (Commons Environment Committee, 1992). The committee recommended that MAFF's responsibilities for coastal defence should be transferred to the then Department of the Environment (DoE) and that the latter department should set up a national unit to oversee coastal zone policies. This would be part of a hierarchy of coastal zone planning from national, through regional, to local level. Corresponding to the publication of the Committee's report the Institution of Civil Engineers held a conference with the theme of integrating coastal zone planning and management (Barrett, 1992). DoE published a discussion document (DoE, 1993) in which the Government's view that there should be integration of economic and environmental considerations in coastal zone management. However, responsibility remained with MAFF.

The importance of the coast as a whole was recognised in the publication of new government guidance in 1992 (DoE, 1992). The introduction to this document states:

> ... it is the role of the planning system to reconcile development requirements with the need to protect, conserve and, where appropriate, improve the landscape, environmental quality, wildlife habitats and recreational opportunities of the coast. This is achieved through development plans and planning decisions, which implement policies for the conservation and improvement of the coastal environment, acknowledging the special character of the coast.

The guidance points out that planning the coast is a strategic issue because the scale over which natural processes operate often spans administrative boundaries and suggests the planning process can be improved by regional conferences or other groupings of local planning authorities to pool knowledge, define key issues and co-ordinate policies. The South East Regional Planning Conference (SERPLAN, 1993) recognised the lowland of the Essex Coast to be of regional significance and identified three elements for the implementation of protection measures:

- Shoreline Management Plans based on sediment cells;
- County Coastal Strategies elaborating on relevant policies in the Structure Plan; and

- Coastal Management Plans generally prepared at District Council level.

Regional Planning Guidance (DoE, South East Regional Office, 1994), written in the context of rising sea levels, said that local authorities should reassess the need for, and location of development in flood risk areas and the implication for sea defences, their maintenance and strengthening, or in appropriate circumstances instigate schemes of managed retreat. It said that this would require local authorities to collaborate in the preparation of coastal policies and to emphasise conservation and, where appropriate, enhancement of the coastal environment.

Attention was also beginning to turn to the estuaries. King (1992) pointed out that existing national policies:

> underestimated the conservation value of 'mud', and thus neglected the importance of estuaries both in wildlife and scenic terms.

Rothwell (1992) emphasised the importance of the estuary environment especially to birds and the threats posed by human interference. He said that designations such as Ramsar Sites, Special Protection Areas, and Sites of Special Scientific Interest were not in themselves a guarantee of protection and pressed for the implementation of adequate management schemes. The Royal Society for the Protection of Birds (RSPB) was also drawing attention to the need for an integrated coastal zone management (RSPB, 1990) and was pressing for there to be a statutory duty on local authorities to prepare Estuary Management Plans. Shortly afterwards, English Nature (1993a; 1993b; 1993c) also promoted the concept of such plans.

## Essex Issues Report

A working group representing ECC and the coastal district councils was set up early in 1993. This produced a consultation draft of issues in late 1993 followed by a final Issues Report in the spring of 1994 (ECC and Essex Coastal Districts, 1994). This classified the issues into physical processes (sea level rise, flood risk, saltmarsh and grazing marsh loss and sea defences), conservation (nature, marine, landscape and archaeology), and development (agriculture, general development pressures, coastal resorts, recreation, aggregates, waste disposal, energy, pollution and fisheries). The report acknowledged the SERPLAN guidelines for three tiers of coastal planning.

### *Shoreline Management Plans*

The Issues Report anticipated that Shoreline Management Plans would be prepared by joint arrangements between local authorities, the National Rivers Authority and MAFF. These would be related to sediment cells and would therefore not correspond neatly to administrative boundaries.

*Coastal strategies*

Coastal strategies would be county based and would elaborate on the strategic land use policies contained in the Structure Plan. The report recognises that the county policies for the coast requires review to meet the statement in PPG20 (DoE, 1992) that Structure Plans provide the opportunity to set out general policies for the coast in consultation with neighbouring counties.

*Coastal Zone Management Plans*

Coastal Zone Management Plans would be prepared at district level. It noted that two such plans were in preparation. The report agreed that these should complement development plans and provide an opportunity to address wider issues that fall outside the town and country planning system. It was noted that English Nature was assisting in the preparation of an Estuary Management Plan for the Blackwater.

**Essex Coastal Strategy**

At the same as the Issues Report was published, a consultation draft of the Essex Coastal Strategy was produced and was followed by a definitive document (ECC, 1994) later in the same year. The landward boundary of the coastal zone was taken to be the same as previous defined in the subject plan (ECC, 1984). As previously, urban areas and other areas where there was commitment for development were excluded.

A major change from what went before was that the boundary did not stop at the shoreline. It was recognised that, although the planning authorities did not have planning jurisdiction beyond the mean low water mark, there was interaction between coastal and marine activities.

The previous policy (NR18) on development, slightly reworded became Policy G2 of the new strategy, and was joined by three other general policies as follows:

- G1. The sustainable use of the coastal zone will be encouraged, and the County Council and coastal District Councils will seek to work towards a reconciliation of conflicts of use along the coastline;
- G2. There shall be the most stringent restrictions on development on the rural and undeveloped coastline outside built up areas. Development, which is exceptionally permitted shall not adversely affect the open and rural character of the coast or its habitats and the wildlife they support;
- G3. The County Council and coastal District Councils will work with other organisations/agencies to achieve better management of the coast. There is a particular need for cross border liaison to ensure a strategic approach to coastal zone planning and management;

- G4. Estuary Management Plans, Shoreline Management Plans, and wider coastal strategies will be supported.

It can be seen from these policies that the aim was a holistic approach reflecting the need for co-operation and co-ordination.

On physical processes, including flooding and coastal protection the following policies are included:

- PP1. Development will not normally be permitted in areas at risk from tidal flooding or coastal erosion where existing defences, properly maintained would not provide an acceptable standard of safety over the lifetime of the development or where new coastal defences would be required. On matters of this type, the local planning authority should consult the NRA for advice;
- PP2 Coastal defences should be maintained where they protect human life and properties, commercial and industrial assets, port facilities and agricultural land, where it can be shown to be economically viable, as well as irreplaceable coastal habitats;
- PP3. Soft Engineering methods of coastal defence will be supported where practicable and appropriate, and appropriate areas for soft engineering should be identified in conjunction with the NRA and MAFF in coastal management plans. Managed Retreat pilot schemes such as those undertaken by MAFF and English Nature will be supported.

PP2 represents an appreciation that defences are no longer to be provided without good economic or environmental justification and PP3 is recognition that alternatives to traditional seawall work, which were considered experimental in the late 1970s and into the 1980s, are now regarded as mainstream.

## Blackwater Estuary Management Plan

Work on this started in 1992 and the plan (Maldon District Council and Colchester Borough Council, 1996) was published in 1996, following the issue of a consultative draft in 1995. The plan has no statutory force and relies on persuasion for its success. Its approach, like that of the strategy document, reflects the realisation that a holistic and co-operative approach was required. A specific statement is made in the introduction to the published plan that its philosophy is:

> to promote the sustainable use of the estuary so that it can yield the greatest benefit to the present population whilst maintaining the potential for the estuary to meet the needs and aspirations of future generations, in a manner compatible with the maintenance of the natural properties of the estuary and its value for wildlife.

In the early stages, Maldon District Council worked with English Nature on the preparation of an integrated management plan for that part of the estuary under its control and from 1995, this collaboration was extended to that part of the estuary falling within Colchester Borough Council's administrative area. Amongst the stated objectives of the plan was:

> to seek a co-ordinated and consistent approach to the planning and management of the estuary as a whole, through the integration of the Management Plan with the relevant statutory Local Plans and Structure Plans and Shoreline and Catchment Management Plans.

The report recognised that consensus and support were essential ingredients of successful implementation and a Statement of Commitment was introduced which included the above objective as the common goal for the plan. This commitment has been signed by a large number of users of the estuary and 'regulatory bodies'.

## Current trends in flood defence

The Blackwater Estuary Plan recognises the importance of saltmarsh as a dissipater of wave energy and notes that seawalls themselves interrupt the natural regime of saltmarsh formation. It argues for 'soft engineering' solutions that work with, rather than oppose natural processes.

As described earlier experimental schemes of this nature began to be considered in the late 1970s, largely as a response to economic considerations. During the 1990s, such work continued but generally gained much wider acceptance, often being considered early in deliberations rather than as a last resort. The early trial of polders at Deal Hall has been followed by schemes at two locations on Mersea Island.

Perhaps more controversially, further schemes of managed retreat have followed the early experiment on Northey Island at Tollesbury Fleet on the north bank and Orplands on the south bank. Dixon et al. (1998) described the multi-disciplinary approach used in assessing the appropriate solution to the Orplands frontage where saltmarsh erosion had made traditional approaches to seawall maintenance unviable. Dixon et al. (1998) also described a scheme of habitat creation at Abbotts Hall Farm on the north bank of Salcott Channel. This differs from the initial managed retreat schemes in that, although the area has been converted to semi-tidal conditions by the opening of sluices originally intended for land drainage and artificial creeks created, the existing seawall has been maintained. It is understood that the success of this scheme is doubtful due to the limited interchange of water possible through the sluice (Hall, 2001). A more ambitious scheme is now proposed for fully managed retreat along a length of 3.5km of seawall (Essex Wildlife Trust, 2000). If successful, this will result in the creation of about 70 hectares.

The work at Orplands and at Abbots Hall has been carried out under MAFF's Habitat creation scheme (MAFF, 1997), which uses EC Common Agriculture Policy funding to encourage environmentally sensitive agricultural practices. Under this scheme work can only be grant aided if the creation of the saltmarsh is consistent with the provision of effective and sustainable coastal defence.

A factor in gaining acceptance of these 'soft engineering' approaches is the attitude of landowners. As indicated in Figure 9.1, there is significant ownership or management by conservation bodies. It may be questioned whether the schemes described would have been acceptable had land remained under traditional ownership (Hall, 2001).

The relationship that developed, largely during the 1990s, is illustrated in Figure 6.4. This may be contrasted with the situation that existed up to the 1970s shown in Figure 6.2.

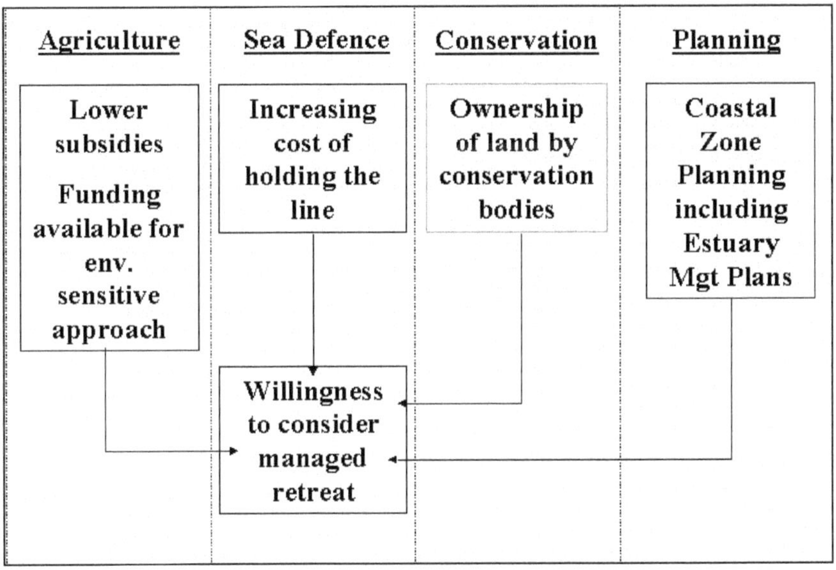


**Figure 6.4 Inter-relationship between sea defence, agriculture, conservation and planning in the 1990s**

### Select Committee recommendations

The Select Committee on Agriculture made a number of recommendations (House of Commons, 1998). Those particularly relevant to the coastal zone are summarised as follows:

- revision of MAFF project appraisal process to allow transparency and adoption of Best Practical Environmental Option (BPEO) approach;
- removal of artificial legal and administrative distinction between flood and coast protection responsibilities;
- regular review of targets set through Shoreline Management Plans;
- pressure to ensure national strategies are translated into practice;

- reorganisation of institutional and administrative responsibilities for flood and coastal defence;
- Shoreline Management Plans to be given statutory status;
- greater effort to develop and implement Coastal Management Plans and to prioritise Shoreline Management Plans within them;
- granting of statutory status to Coastal Groups.

The need for strengthening the attention paid to habitats in SMPs had been pointed out by, amongst others, Friends of the Earth (1997). Government response to the Select Committee recommendations was mixed (HM Government, 1998). It pointed out that the system of Project Appraisal Guidance Notes were in the process of revision and would address the Select Committee's recommendation on BPEO. It was not keen to change any administrative arrangements. On translation of national strategy into practice, the Government made specific reference to the Orplands scheme as an example of what could be achieved. The Government's approach to flood and coastal defence were reiterated during a House of Lords debate (Hansard, 2000).

## Coastal Habitat Management Plans

In the Spring of 2000, it was announced (MAFF, 2000) that the Essex Coast and Estuaries would be one of 6 initial areas for which Coastal Habitat Management Plans (ChaMPs) would be prepared and that funding for subsequent projects would come from MAFF's flood defence budget. It is perhaps significant that the Countryside Minister chose to launch the EU funded scheme at Abbotts Hall.

## European developments

A report to the European Commission examined how the aim of Integrated Coastal Management (ICZM) could be best achieved throughout the European Union (Gibson, 1999). The report notes out that there is a frequent distinction in existing laws between those governing sea and land and that powers, such as planning control, given to local authorities are often confined to land based development and that there is therefore difficulty in adopting a holistic approach to coastal zone management. It stops short of recommending a EC Directive for ICZM.

The lack of co-ordination is also acknowledged in a strategy document by the Commission itself (Commission of the European Communities, 2000). Its proposal states:

> The Strategy aims to promote a collaborative approach to planning and management of the coastal zone, within a philosophy of governance by partnership with civil society.

The strategy sees the EU role as to provide leadership and guidance by establishing a framework to enable action at other levels. The introduction to the Principles of ICZM contained in the Annex to the Strategy states that ICZM is:

> a dynamic, multi-disciplinary and iterative process to promote sustainable management of coastal zones. It covers the full cycle of information collection, planning (in its broadest sense), decision-making, management and monitoring of implementation. ICZM uses the informed participation and co-operation of all stakeholders to assess the societal goals in a given coastal area, and to take actions towards meeting these objectives. ICZM seeks, over the long-term, to balance environmental, economic, social, cultural and recreational objectives, all within the limits set by natural dynamics.

## Conclusion

Management of seawalls in the Blackwater Estuary has evolved from a policy of fighting nature so as to 'improve' agricultural productivity to a situation where the importance of working with nature is recognised. This evolution has been brought about by changes in national and local policies with regard to coastal management, in particular the recognition that an integrated approach is needed. The evolution in approach has been encouraged and aided by the conservation bodies who have become significant landowners along the estuary. It can be conjectured that, had such changes in land ownership not occurred, the application of the new approach would have been stifled. Whilst not perfect, the approach used on the Blackwater has the potential to act as a demonstration, both nationally and internationally, of what could be achieved.

## Acknowledgement

I would like to thank Nicky Spurr (Blackwater Project Officer at Maldon District Council) and John Hall (Director of Essex Wildlife Trust) for their helpful comments on early drafts of this chapter.

## References

Barrett, M.G., 1992. *Coastal Zone Planning and Management*, Thomas Telford Ltd, London.
CEC, 2000. *Integrated Coastal Zone Management: A Strategy for Europe Communication*, Commission of the EC, Brussels, COM(2000) 547 final.
Commons Environment Committee, 1992. *Coastal Zone Protection and Planning*, HMSO, London.
Department of the Environment, 1972. *Planning of the Undeveloped Coast*, HMSO, London.
Department of the Environment, 1992. *Planning Policy Guidance PPG20 Coastal Planning*, HMSO, London.
Department of the Environment, 1993. *Managing the Coast*, HMSO, London.
Department of the Environment, South East Regional Office, 1994. *RPG9: Regional Planning Guidance for the South East*. Available from http://www/gn.apc.org/pmhp/dc/planning/rpg9.htm. (Accessed 9th November 2000).
Dixon, A.M., Leggett, D.J. and Weight, R.C., 1998. Habitat creation opportunities for landward coastal re-alignment: Essex case studies, *Journal of Chartered Institution of Water and Environmental Management*, **12**(2), 107-112.
English Nature, 1993(a). *Estuary Management Plans – A Co-ordinators Guide*, English Nature, Peterborough.
English Nature, 1993(b). *Important Areas for Marine Wildlife around England*, English Nature, Peterborough.
English Nature, 1993(c). *Strategy for the Sustainable Use of England's Estuaries*, English Nature, Peterborough.
Essex County Council and Essex Coastal Districts, 1994. *The Essex Coast: Issues Report*, ECC, Chelmsford.
Essex County Council, 1966. *Interim Coastal Protection Policy*, ECC, Chelmsford.
Essex County Council, 1982. *Approved Structure Plan 1977-1991*, ECC, Chelmsford.
Essex County Council, 1984. *Essex Coast Protection Subject Plan: Written Statement and Proposals Map*, ECC, Chelmsford.
Essex County Council, 1991. *Essex Structure Plan First Alteration 1986-2001*, ECC, Chelmsford.
Essex County Council, 1994. *Essex Coastal Strategy*, ECC, Chelmsford.
Essex Wildlife Trust, 2000. Abbotts Hall Farm and Joan Elliot Visitor Centre, *Essex Wildlife*, **52**, 10.
Fleming, C.A., 1989. The Anglian sea defence management study. In: Institution of Civil Engineers, *Coastal Management: Proceedings of Conference*, 153-164, Thomas Telford Ltd, London.
Friends of the Earth, 1997. *Briefing Paper: Lost at Sea? – Shoreline Management Plans and Coastal Defences*, FoE.
Gibson, J., 1999. *Legal and Regulatory Bodies: Appropriateness to Integrated Coastal Management*, Report to European Commission DGXI by MacAlister Elliott and Partners Ltd, Lymington.
Grieves, H., 1959. *The Great Tide: the Story of the 1953 Flood Disaster in Essex*, Essex County Council, Chelmsford.
Hall, J. (Director of Essex Wildlife Trust), 2001. Personal communication.
Hansard, 2000. House of Lords Debate for 7th June 2000, Columns 1157-1189.

HM Government, 1998. Government response to the agriculture select committee report on flood and coastal defence. In: House of Commons, 1998. *Select Committee on Agriculture 5th Special Report*, The Stationery Office, London.

House of Commons, 1998. *Select Committee on Agriculture 5th Special Report*, The Stationery Office, London.

Kamps, L.F., 1962. Mud distribution and land reclamation in the Eastern Wadden Shallows, *Rijkswaterstaat Communication*, **4**.

King, G.A.D., 1992. Setting the scene: the planning dimension. In: Barrett, M.G. (eds.), *Coastal Zone Planning and Management*, Thomas Telford Ltd, London.

Leggett, D.J. and Dixon, A.M., 1994. Management of the Essex saltmarshes for flood defence. In: Falconer, R. and Goodwin, P. (eds.), *Wetland Management*, 232-245, Thomas Telford Ltd, London.

Maldon District Council and Colchester Borough Council, 1996. *Blackwater Estuary Management Plan*, MDC, Maldon.

Ministry of Agriculture, Fisheries and Food, 1997. *The Habitat Scheme: Saltmarsh – Guidelines for Farmers*, MAFF Publications, London.

Ministry of Agriculture, Fisheries and Food, 2000. *Coastal Wetlands to be Saved*, MAFF News Release 3rd April 2000, Available at http://www.maff.gov.uk/inf/newsrel/2000/000403a.htm (Accessed 9$^{th}$ November 2000).

Ministry of Housing and Local Government, 1963. *Coastal Preservation and Development*, MHLG, London (Circular 56/63).

Ministry of Housing and Local Government, 1966. *The Coast*, MHLG, London (Circular 7/66).

Rothwell, P.I., 1992. Estuaries. In: Barrett, M.G. (ed.), *Coastal Zone Planning and Management*, 101-111, Thomas Telford Ltd, London.

Royal Society for the Protection of Birds, 1990. *Turning the Tide – A Future for Estuaries*, RSPB, Sandy.

South East Regional Planning Conference, 1993. *Coastal Planning Guidance for the South East*, SERPLAN, London.

Chapter 7

# Characteristics of Land-Related Administrative Institutions: What is Needed to Support Equitable and Productive Land Development?

David Bledsoe

## Introduction

Although land-related administrative institutions vary from country to country, they are typically intended to fulfil similar roles and provide similar services. Some of the functional areas within which they operate include land titling and registration, land use planning, control, and monitoring, dispute resolution, valuation and taxation, surveying and mapping, credit and mortgage, insurance, agency and brokerage.

Characteristics such as the political system, maturity, and activity of the land market, levels of foreign and domestic investment, per capita income, historical mapping, and cadastre efforts, together with the structure of the taxation system influence the relative distribution of supporting resources between the functions. Developing and transition countries may offer few of these functions. One estimate of the distribution of public sector resources dedicated to land-related functions in Central and Eastern Europe (CEE) is illustrated in Table 7.1.

**Table 7.1 Distribution of public sector resources in CEE**

| | |
|---|---|
| Public sector estate management | 25% |
| Valuation, including property tax assessment | 10% |
| Physical planning and building control | 25% |
| Surveying and mapping | 5% |
| Land registration | 25% |

*Source:* Based on FAO, 2002, Guidelines on Rural Property Tax Systems in Central and Eastern Europe

Not surprisingly, transition countries tend to dedicate more resources to public land management because of the higher proportion of publicly owned land.

To whatever extent they exist to fulfil these functions, the land-related institutional framework should rationally develop and clearly define the rights and obligations of both the land rights holders and the administrative actors, and provide mechanisms for changing the rights and obligations. Furthermore, sustainable administrative processes should be created that permit and nurture efficient transactions, and provide administrative mechanisms for resolving disputes in an equitable but expedient and efficient way, and for promoting legal literacy. Finally, the framework should be sensitive and responsive to often neglected groups, such as women, minorities, and the poor, who are frequently excluded by land institutions. If the institutions fail, productive and efficient land use and land transactions are hampered or simply fail to happen and the anticipated benefits either fail to materialize or are stunted and not worth the cost. Examples provided later in this chapter show that institutional inadequacy, deformity, or absence can interfere with or prevent beneficial land transactions and uses.

## Institutional place and position

### *Distribution of institutional responsibilities and functions*

The rational allocation and distribution of responsibilities and functions amongst and across levels of land-related administrative institutions are essential to accomplish institutional missions. Several perspectives are at issue. One is about where land policy is made and implemented. Certain fundamental principles, such as the private ownership of land, including the right to buy and sell, are best laid out at the national level. At the same time, some related, subsidiary rules (such as the maximum permissible size of agricultural land holdings) might be developed locally (but perhaps within a range of possible choices that has been set nationally).

The aim of central government bodies should be to provide the general policy and regulatory mandates from which regional and local administrative institutions can operate. The aim of local institutions and the laws defining their authority in land-related matters should be to efficiently and transparently implement the general policy framework in the light of local information and characteristics. Under any circumstance, crafting policy at the local level would assuredly lead to disparate approaches and results. For example, a national government might appropriately conclude as a general policy, that a viable credit and mortgage market would benefit the agricultural sector. Mortgage regulations and standard mortgage forms and contracts might also appropriately be crafted at the national level. However, mortgage filing and foreclosure actions would probably be best conducted at the local level (Giovarelli, 2000). Land titling and registration should probably be standardized at a national level, but the acceptance and processing of registration documents from rights holders ought to be done at the local level, to efficiently capture transactions and to minimize transaction costs for those registering rights. A centralised registry for all system information might then be kept at a central level. To take another example, general tax policies might best be crafted at a national or regional level, whereas tax valuations might be administered most efficiently at a local level (FAO, 2002). Finally, local land use or zoning plans are also probably best crafted (and variances from them granted) at the local level, but such local plans should reflect the general policy mandates provided by the national or regional government (Prosterman and Hanstad, 1999).

Concentration of responsibility is yet another perspective. How many roles or tasks should be assigned to an institution? Capacity should be considered when distributing responsibility and authority to institutions in enabling legislation. Numbers of staff, office size, and distribution, together with funding will affect assignments of responsibility. Equity should also be considered. Placing too many areas of authority within a single institution may create conflicts of interest and the prospect for abuse of power. Some combinations of functions may best be distributed to more than one institution. For example, in land use regulation, the functions of planning, regulation promulgation, permit application processing, variance rulings, inspection, and sanctioning might best be split between two or more bodies.

However, fragmentation of responsibility can also become an issue affecting the distribution of authority, particularly in evolving governments. This occurs when discrete or closely interrelated functions are distributed across a number of institutions, either vertically or horizontally. In Albania, a lack of communication and coordination between the five ministries collaborating in a new land registration system created confusion and duplication of efforts (Stanfield, 1996). In Bolivia, a large number and confusing framework of government agencies assumed responsibility for titling land, which created the prospect of duplicate titles and uncertainty about rights (Barnes et al., 2000). Fragmentation at any level can create confusion, inefficiency, and poor or non-existent services. Fragmentation can occur when old institutions remain in place after their functions have changed or been eliminated. As new policies and related functions are created, they may be pushed upon a collection of old institutions with resulting inefficiencies. With the creation of new functions there should also be an evaluation of the capacity of existing institutions to undertake them and, if necessary, old institutions should be eliminated. If new functions are distinct from present functions, new institutions may be created to operate alongside old institutions. In other cases, fragmented institutions should be combined.

Finally, the simple failure to assign administrative responsibilities can lead to unintended and negative results. When the Kyrgyz Land Code was passed, no specific agency was named to administer the sweeping reforms. Before the reforms, land had been administered by several agencies. A power vacuum and struggle subsequently resulted. The strongest agency, now called GosRegister, emerged to fill the void, but has been seen to abuse its power at the expense of land rights holders. For example, despite legislative assurances that land plots will be distributed and formalized free, some local GosRegister officials require payment before they will survey or register land and, furthermore, there is no overseeing agency (Giovarelli, 2002).

*Decentralisation*

As mentioned above, some responsibilities and powers are best left with centralised (national or regional) institutions whilst others are better given to local institutions and governments (*see* Fox, 1994; World Bank, 2001; World Bank, 2002). From the standpoint of administrative efficiency and empowerment, institutions such as the European Union (EU), the World Bank, and the U.S. Agency for International Development (USAID) have urged the principle of pushing functions to the lowest administrative or operational level at which there is a clear capacity to carry them out (Hennessy, 1998; World Bank, 2001; USAID, 2002).

Thus, the issue of institutional decentralisation is usually seen as one of moving responsibilities and authority to local levels. However, decentralisation can be viewed in at least two ways: firstly by moving central government offices to regional or local levels (deconcentration), and secondly, by moving responsibilities and authority from the central government to local levels (devolution). Deconcentration can be done without moving authority to local levels from the centre. Devolution is primarily what is being promoted by the EU, World Bank, and USAID.

There are dangers at both ends of the continuum. Over-centralisation of land-related functions can cause confusion, delay, and inefficiency (Farvacque and McAuslan, 1992). For example, over-centralisation of some land administration functions restricts land use changes in Russia and the Kyrgyz Republic (Prosterman and Hanstad, 1999). In Guatemala, over-centralisation and the reality of only two registry offices serving the entire country made it difficult or impossible for many to register their parcels (Barnes et al., 2000). During later stages of land reform in Zimbabwe, problems that stemmed from over-centralisation of government, application of top-down methods of planning and implementation, and poor communication among ministries, in conjunction with a more general lack of incentives, created resistance and damaged the popularity and progress of the reform process (Alexander, 2000).

Decentralisation carries risks with it as well. For example, decentralisation of land-related functions without accompanying, training, resources, or support from the centre has created a local administrative and regulatory void in many parts of Uganda (Muhereza and Bledsoe, 2002). In some places within the Kyrgyz Republic, decentralisation of the power to lease state-owned land fund parcels, and a lack of accompanying monitoring and mechanisms for accountability, has resulted in favouritism and inequitable distributions (Giovarelli, 2002).

It is clear that the process of decentralisation of land-related administrative functions must be careful and only made after appropriate evaluation and planning. Fox (1994) makes an informative analysis of the process of evaluating the decentralisation of infrastructure functions. In this context, devolution works well when economies of scale are limited, local input and choice is strongly desired, demands vary by locality, benefits from the institutional service delivery accrue to local residents, and strong local government exists. Devolution is less effective when many beneficiaries are distributed across a wide geographic area, demands for service are homogeneous across localities, and economies of scale are significant.

Although infrastructure services are much more capital- and plant-intensive than land-related services, some of these criteria might be useful. For example, in the developing country setting, the evaluator might reasonably conclude that titling and registration services might be appropriate at a regional level rather than at the lowest local level. This is because there are some economies of scale, local input and choice is not much desired, demands do not (or ought not) vary much by locality (other than in the urban/rural context) and tend to be homogeneous across localities, strong local government typically does not exist, and many beneficiaries are distributed across a wide geographic area.

Another way of perhaps analysing the feasibility of decentralisation is to determine whether the requirements for success are in place or can be created. The degree of transfer might reasonably depend upon how closely the requirements are met. Local institutions need to have skills in decision making, administration, and the appropriate functional areas. A lack of skills means that costly training will be required in many cases. In addition, rural labour markets may not attract or retain skilled workers. The characteristics of other local institutions must also be considered when shifting responsibilities and authority to local levels. For example, shifting land registration duties to a local level may be problematic if there are no local surveyors. Central government support may be required in such cases. Local institutions should reflect local customs and norms such that the institutions are seen as legitimate and fair. If local support is not secured, the costs of obtaining and enforcing compliance can be prohibitively high. The local institutions should be reasonably free of corruption and favouritism. Local institutions must be enabled to raise and retain revenue, while simultaneously being able to depend on predictable grants from central government (Fox, 1994; Baker, 1994; Farvacque, 1992).

Using a similar approach, the World Bank attempts to analyse the context and capacity of the lowest organisational level of government that will be responsible for carrying out functions (World Bank, 2002). The Bank also perceives the need for the connection of local financing and fiscal authority to the devolved responsibilities. Local community education and participation is also viewed as being of great importance together with the development of a binding mechanism that the community can use to express its preferences to politicians, to create a credible incentive for people to participate, and a level of accountability and monitoring so that politicians and local officials have an incentive to be responsive.

Importantly, the underlying goal should be to attain the proper balance of centralised and decentralised systems. National governments have a key role in creating the conditions that will enable local level governments to carry out their responsibilities. The success of decentralisation is often contingent upon the extent and quality of training for all of those involved in decentralisation administration (World Bank, 2002).

The factors weighing against devolution and the challenge of meeting the requirements for success may prompt some compromise between leaving institutional responsibility and authority in centralised hands and moving it to rural local governments. As suggested above, regional institutions are one option. Co-operation and collaboration between local institutions in some function, and the application of new technologies to deliver quality service to sparse populations are other options (Brown and Reid, 1990).

## External influences

Many external factors influence land-related institutions, including international donors, local organisations, and NGOs (Agarwal and Ostrom, 2001). Also included are government agencies and the private sector (UNECE 1996). This section examines some of these influences, and suggests ways in which institutions can respond in an appropriate manner.

### *Trends in theory*

Varieties of new and evolving theories have worked to keep land-related administrative institutions in flux. For example, New Public Management (NPM), a public organisational/administration theory that has been promoted and embraced by many around the world, emphasises decentralisation, devolution, and the modernisation of public service delivery. NPM is generally seen to carry with it organisational and managerial practices born of private business and the marketplace. The government austerity programmes and structural adjustments promoted by the International Monetary Fund and the World Bank incorporate such marketisation principles as NPM. Some believe that a problem can arise when the state, acting in some ways as a private service provider, loses sight of and fails to uphold or promote the public interest. In a land context, this shift could conceivably include charging unrealistically high fees for services, thereby limiting their use and excluding some disadvantaged groups. Emphasis is frequently placed upon monitoring and reporting procedures that focus on quantity to the exclusion of quality. In some instances, responsibilities are placed at levels that do not have the capacity to accomplish them. Finally, there is a trend to reduce staff numbers below the levels needed to start and perpetuate fledgling institutions.

Conversely, private business principles and the privatisation of services in some cases might result in improvements to the quality and efficiency of institutional product. Privatisation of certain services may present a realistic solution to corruption-wrought agencies, as private sector accounting methods and the need to operate for profit create strong incentives against corrupt practices (Transparency International, 2000; Smith, 2001). Privatisation of functions should be viewed as a continuum of options running from purely private service provision to a variety of public/private partnerships. When evaluating the functions of public institutions for possible privatisation a number of criteria are pertinent, including:

- competitive market pressures must exist in the area to be considered for privatisation. For example, if surveying or computer system support were to be considered for privatisation, contestable markets (markets where potential entrants restrain the price-setting behaviour of current private producers) in surveying and computer support must exist or be created;

- privatisation works best for services that are already delivered in the private sector;
- services that have easily observable output and that can be easily monitored are best suited to privatisation; and
- services that may be subject to operational efficiencies are better candidates for privatisation (Fox, 1994).

As another example, the issue of fees for service is not easily resolved. Although the majority of funding for developing-country, land-related institutions now typically comes through tax revenue subsidies, some services provided by such institutions might appropriately be paid for (at least in part) by the users. Land title registration and recording services are an example. Most developed countries have instituted a relatively low registration fee that partially or fully covers the costs associated with land registration. A caveat here is that when public policy requires motivating persons to use the service (such as land registration) to ensure broad coverage, the level of fee should not discourage use. Targeting subsidies toward those service costs can ensure access for the lowest income service users. Furthermore, payment of fees should be clearly linked to receipt of services. Service enhancements can be accompanied by appropriate fee increases although imposing service fees can be politically difficult initially in areas where services have been under-priced or fully subsidised. However, the failure to price services can create problems if revenue shortfalls lead to dissatisfied users and slower economic growth. Other candidate services for user fees might be surveying, mapping, development permit processing, and zoning variance requests.

At the risk of oversimplification, the answer is to adopt only those measures that make sense under the circumstances: efficiency and the public good must be balanced. An awareness of the options and their effects is the first step. Thorough evaluation (to the extent possible, given limited resources) and reasonable but effective implementation must follow.

Other theories are related to land, and the clarity of hindsight should direct us to use the same approach for land-related trends. The overall missions of land-related institutions might be affected and parcel individualisation and privatisation might serve as broad examples. When examining at the application of land titling and registration of individual parcels in parts of Africa, the consensus seems to be that it has not spurred economic growth or alleviated poverty, and that it probably has caused some harm (*see* Bruce, 1986; Firmin-Sellers and Sellers, 1999; Quan, 2000; Platteau, 2000). First, individualisation and titling of communal lands can create landlessness when common users are left without common land resources (Bruce, 1999; Muhereza and Bledsoe, 2002). Second, there is little or no evidence that land titling and registration of individual parcels has spurred investment or consequently worked to increase agricultural production in Africa. In fact, evidence from Kenya and elsewhere indicates that it has not increased investment

(Quan, 2000). In addition, observations from Uganda indicate that some specific forms of tenure insecurity may, conversely, spur investment Parcel occupiers may increase investment in perennial crops and other improvements so as to make eviction more costly by virtue of higher mandated compensation amounts (where compensation requirements are enforced) (Mackinnon and Reinikka, 2000). Third, there is little evidence that titling and registration has worked to increase opportunities to obtain credit in Africa. In Kenya, smallholders avoid applying for loans secured by their land because they are afraid of foreclosure. In Uganda, landholders are also presumed to be fearful of using secured credit because of a fear of foreclosure. Finally, again supported by evidence from Kenya, it appears that titling has not prompted an increase in land market transactions or in transfers that display evidence of increased efficiency of land use.

The lesson is that sufficient information must inform the analysis of feasibility of projects and the consequent design of institutions. The information must include a picture of traditional practices and customs, the potential for successfully importing new practices and economic frameworks, and realistic appraisals of the capacity of governments to implement selected schemes.

### *Technological changes*

Major technological changes usually come from outside a developing country and information technology improvements are a frequently encountered example. The computerisation of land registries has been one of the most important technological changes over the last decade, but ideas on other ways to use new technologies have also been evolving (Dale and McLaughlin, 1999). In some contexts, problems may arise if inappropriate technologies are introduced. In Albania for example, while records related to initial parcel registration were being computerised at the centre, no capacity for computerising subsequent transactions was developed at the local level. The integrity of the central database was quickly eroded. Also in Albania, computer-generated base mapping was used centrally to plot initial registration of parcels. At the local level, changes in parcels, such as sub-division, are being recorded by hand on paper copies since there was no capacity for computer mapping at the local level, and no system to route changes to the centre was put in place (Gaynor and Bledsoe, 2000). However, the benefits of using technologies that are well suited to individual contexts are numerous. To confirm suitability, all technologies should be justified in terms of speed, cost, installation, training and personnel, maintainability, accuracy, cultural and social applicability, and practicality and simplicity of field operations (Barnes and Eckl, 1996; UNECE 1996).

### *International bodies*

A variety of private, quasi-public, and public international entities affect land-related administrative institutions. The development banks provide a good example, providing concessionary loans and the grants for projects, and defining conditions and frameworks for how institutions should be structured and operated.

Programmatic philosophies and objectives are born of the needs and perspectives, held by these lenders and loan repayment becomes a major consideration. Land-related efforts are conceived of, in part, by development banks and serve as targets for loan funds. These include parcel privatisation and individualisation, systematic demarcation (including mapping and surveying), tax systems, land use planning and zoning, systematic titling and registration. The supporting institutions that are created as a part of these programmes are shaped to reflect the objectives of the programmatic philosophy.

International aid donors also deliver programmatic form and substance along with their aid funding. Project designs reflect the economic, political, and institutional philosophies of the donor country and, in some cases, insist upon the use of equipment and expertise of that country. For example, the democracy and governance projects of USAID typically include a component that focuses on institutional decentralisation (USAID, 2000). Furthermore, technical assistance is frequently tied to donor-country consultants, which can introduce donor-country biases.

International professional organisations involved in land administration (surveyors, appraisers and valuers, brokers and agents, cartographers, and cadastral specialists, for example) play an important and influential role in helping to develop strategies and provide evaluation in guiding the overarching development of land-administration systems (Robertson, 2002). Many of the principals and members of these organisations also serve as consultants. Information and approaches shared between and developed by these experts are then often implemented in developing countries.

Quasi-public entities also have a role in land and the related institutional issues. For example, the UN Human Settlement Programme recently launched the 'Global Campaign for Secure Tenure', which focuses on the role of land tenure in achieving sustainable development. The UN Development Programme has funded a variety of land-related projects around the world. The Food and Agriculture Organisation of the UN also funds a variety of land-related projects, as does the related International Fund for Agricultural Development. All of these organisations bring to their land projects the impress of their missions and their organisational objectives.

Finally, international NGOs play a role in shaping land-related institutions and the ways in which they operate. The list is long, and focused NGOs dealing with gender, indigenous peoples, sustainability, community participation, civil rights, democracy, and other topics all seek to influence land and the administering institutions. In addition, more broadly focused NGOs dealing with human rights, poverty, and the like, are constantly attempting to further their goals as they relate to land.

Given the volume of advice, counsel, approaches, consultations, philosophies, and conditions that are made available or imposed upon the land institutions of developing countries, challenges are created simultaneously. First, the developing country counterparts must do the best that they can (given limited resources and alternative sources of information) to evaluate the philosophies and approaches that accompany concessionary loans, donor funds, or technical assistance. In-country experts and academics should be consulted and used as partners as much as possible. However, the outside information should not be simply dismissed as being biased or directive since much of it is useful and contextually appropriate. Second, developing countries need to be able to reject or modify programme components that do not mesh with their goals, traditions, and perceived needs, without fear of jeopardising the offer of assistance. Third, lender and donor representatives need to view programme design as a consultative and collaborative effort. Target country representatives are sometimes viewed as the opposition or persons in need of enlightenment. Importantly and commendably, recent publications from the World Bank, AusAID, and USAID recommend cautions about, and approaches to, obtaining in-country counterpart input and consensus (World Bank, 2002a; AusAID, 2000; USAID, 2000).

*National trends and actors*

National policies and politics will naturally come into play in shaping and influencing land-related institutions. Broad policies about governance, civil society, resource use, the rule of law, etc. will directly influence the way institutions operate. Furthermore, because land issues are concerned with the distribution of life sustaining resources, they are deeply connected to and inseparable from political issues. As such, political pressures, compromises, and conflicts influence the creation and operation of administrative institutions. To offset this reality, administrative institutional authority should be accompanied by freedom from political intervention. This is more likely if they have secure sources of funding, if politicians protect senior personnel from unfair dismissal, and if senior personnel are not in a position to benefit from the political process. Freedom from political intervention can be fostered if the enabling legislation of an institution sets out key powers and duties, and establishes financial security (Asian Development Bank, 1998).

The Civil Service in developing countries is larger, less productive, and costly given the delivery and production levels. At times, the main motivation for the existence of such a large bureaucracy is frequently to create and redistribute rents to favoured groups of the population. Another motivation is that government jobs are especially appealing and useful in countries that have high levels of external risk. When a government provides numerous secure jobs in the public sector, the income and consumption risk being faced by households in that country drops. Although the immediate benefits are experienced by those working in government positions, additional benefits may reaped by extended family members (Rodrik, 2000).

*Local trends and actors*

Local effects have perhaps the greatest influence on administrative institutions since most implementation is carried out locally. Many local traditions and social practices have a significant influence in determining whether the practices of administrative institutions will meet desired objectives and to deliver sought-after services. In some parts of Uganda, for example, women are not permitted to attend and participate in agricultural extension efforts. Since no efforts are made by the extension services to contact women, they go without needed training and support (Eilor and Giovarelli, 2002).

Institutions created by the state may not link with the informal rules and customs that influence individual and social group behaviour. These traditions and social practices, through both individuals and informal institutions, must be taken into consideration if they are to be accepted and implemented. Furthermore, visible and non-visible hierarchies exist within every social group and locality and it is important to identify their existence and determine how these social-power relations may affect institutions. In Ethiopia, for example, administrators failed to generate partnerships with those in local traditional institutions, and the result was that the land resources project met with little success (Admassie, 2000). Traditional institutions must not be rejected and, instead, policymakers and administrators need to find ways to incorporate or respectfully respond to traditions and social practices. In parts of Africa, many new land tenure and related administrative models now being proposed are based on adaptation to existing traditional systems, as opposed to earlier models that advocated the replacement of customary regimes.

Local NGOs also influence administrative institutions and, in some cases, they are considered more accountable to their constituencies than official or administrative structures, and may serve as surrogates for ineffectual institutions. In some parts of Bangladesh, for example, NGOs have come to replace local, faltering state institutions (Jreisat, 2002). In addition, NGOs can form a natural link between institutions and society, and in the extreme can come to embody the civil society that acts as the collective intermediary between society and the state (Pearce, 2000).

## Internal characteristics

Internal realities define institutions and dictate whether they achieve what they are supposed to. This section examines some of these characteristics, describing them and then suggesting how the better ones can be created and retained, and how the poor ones can be eliminated.

### *Operational and production processes*

The functional processes created and undertaken by land-related administrative institutions must be clear and defined in a rational manner to reflect their enabling mandate that has been established by law or regulation at higher levels of government. Processes should also reflect local characteristics and needs. Definitions, especially for assigned duties and procedures to obtain services, should be clear so that evaluation by outsiders is feasible (Transparency International, 2000). Failure to define processes and to identify their tasks leads to inconsistency, arbitrary actions, duplication of effort, and opportunity for corrupt practices. Furthermore, institutional processes at any level should match fiscal realities and resources. The administrative and financial replicability of processes (for example, methods of land surveying used for title documents) over the entire jurisdiction must be a central consideration (Prosterman and Riedinger, 1987), and regular reassessment and redefinition of processes should be a part of each institution's charter.

National laws related to institutional operations, and the regulations and procedures that are crafted by the institutions themselves, must be well drafted, consistent, coherent, and able to be followed. Bad laws and regulations can hamper institutional operations, land transactions, and uses. For example, restrictive land use regulations that provide overly broad administrative discretion, accompanied by harsh sanctions, stifle efficient agricultural land use in Russia, Belarus, Bulgaria, Kazakhstan, Slovenia, Uzbekistan, Moldova, Ukraine, and the Kyrgyz Republic (Prosterman and Hanstad, 1999). In many Latin American countries, lawyers are often needed to interpret the verbose and confusing documents related to land rights and transactions (Barnes et al., 2000). Vague rules on local government spending have created confusion and inefficiency in distributing land tax revenues in Bulgaria, Hungary, and Poland, whilst in Romania and Bulgaria limitations on the discretion of local government spending have deterred effective use of the revenue from land tax (Prosterman and Hanstad, 1999). Rule-making should reflect national policy while embracing local characteristics and needs and local flexibility must be balanced against the danger of local caprice and arbitrariness within each country-specific setting, and with an eye to the capability, extent of corruption, and degree of ideological bias likely to be present in the local administrative apparatus

*Transparency*

Institutional transparency can be divided into four types: structural transparency, functional transparency, transparency of input, and regulatory transparency.

Structural transparency relates to the institutional composition and organization. Hierarchy, named functions, locations, access, position descriptions, and the names of personnel should be published and available. This availability allows for transparency of relationships, which is necessary to deter potentially corrupt relationships (Transparency International, 2000). Facilities and offices should be open to the public, and tours and open houses should be held periodically. Ideally, there should be one clear point of access to any system for the public, meaning that all information could be accessed at one conveniently located facility, in one visit (Dale and McLaughlin, 1999). This may call for a highly trained team of staff, in a broad arena of information, or that information is printed in a form that is easily understood. In many countries, this may mean that illustrations are provided for those who are illiterate.

Functional transparency requires that the responsibility, decision-making role, and function of the institution are clear and fully understood. The public must either know or be able to understand the scope of institutional responsibility. This requires focusing on the implementation of client-driven goals, which should be assessed by asking; who are the institution's ultimate clients, what are the needs of the clients, and how best they can be satisfied. To develop a formal relationship with the clients, there must be an aggressive educational outreach agenda in addition to transparent institutional functions. Failing this, a large portion of the society will continue to function within an informal system and the goal of securing rights to land will be circumvented (Dale and McLaughlin, 1999). Re-assessment and periodic updating of the institution's relations with the targeted clientele are essential and should include not only a top-down assessment from within the institution but also a public participation element.

Transparent input and operating data mean that the information collected, processed, and acted upon by the institution (e.g., tax base and valuation data, or land registration records) is accessible and easy to understand by citizens who need it to make decisions. Incomplete, outdated, or poorly organised input data creates uncertainty and risk for both the institution and users (Dale and McLaughlin, 1999). For example, a country or region could establish either an active or passive land registry, an active registry meaning that the data being input is actively checked for correctness, whilst a passive registry would mean that the data is simply input and stored but not checked. Some techniques for quality control of data include periodic sample checks (which may include randomly selecting a registered parcel and updating the title information or physically comparing the land to the description in the record) and the development of a professional body to regulate private or even public service providers (such as a professional association of surveyors). As discussed above in relation to transparent regulations, the transaction costs for clients using any land-related institutional system must be kept to a reasonably low level. For example, the client population, especially in developing countries, must believe that the benefits of keeping land registration current outweigh any costs.

Regulatory transparency means that applicable legislation, charters, codes, laws, and interpretive decisions are publicly available together with suitable summaries, checklists, guides, forms, and directories. Institutional staff should be trained and available to locate and interpret obscure regulations.

*Sensitivity and accessibility to disadvantaged groups*

Land administration institutions are not accessible to everyone. Women and others who have a lower social or economic status often face unique obstacles that prevent them from accessing or using the services and benefits. In some cases, the institutions are legally biased against groups of people. This bias is often the result of laws that do not establish or protect the rights of women and others who belong to socially marginalised groups (Agarwal, 1998). In other cases, however, laws mandate that women and others are entitled to equal rights to land tenure (Ericsson, 1999). However, despite these laws, biases remain and many people are denied access to the benefits offered by the institutions (Agarwal, 1994; Deere and Leon, 2001). Although it is not just women that face problems of access and equity, this section primarily addresses problems of access for women. Most of the information, analysis, and recommendations provided here can be extended to other excluded or disenfranchised groups.

The laws governing legal titling and registration of land parcels can leave women without legal rights to the land and often fail to recognise women as legitimate owners, since social perspectives portray men as heads of households and women as dependents. At times, a process of registration pursuant to these laws may legally supersede and negate de facto rights that existed prior to the implementation of land reforms (Barry and Fourie, 2002).

Even when the law provides for women's legal rights to land, institutions can fail to formalise them. In Kenya, Uganda, for example, current laws technically allow women to own land, but the majority of it has been registered in men's names (Munyakho, 1994; Eilor and Giovarelli, 2002). In some Latin American countries, women with legal rights have been excluded as beneficiaries because registration processes failed to recognise them as having legitimate individual rights (Deere and Leon, 2001). Evidence from South Asia also provides examples where men have been recognised as household heads that have legitimate rights, while women lack those same rights (Agarwal, 1998).

When land-related administrative institutions fail to acknowledge formal legal rights they are usually complying with disadvantaging customary inheritance, marriage, and divorce laws. When this happens, those that monitor the performance of the institutions need to determine whether the persistence of customary practices stems from an implementation failure or from poorly crafted legislation (or both). In the former instance, institutional corrections need to be made related to product delivery, whilst in the latter, additional information needs to be gathered about the customary practices and the legislation or implementing regulations need to be amended to bring them into some kind of conformance with the customary realities.

At other times, cultural expectations of household roles and responsibilities constrict women's access to land-related institutions (USAID, 1997). For example, although women frequently provide much or most of the agricultural labour, they are often not taken seriously as agriculturists since farming is often seen as a male activity. Thus, women can be isolated from contact with land-related agricultural institutions. In addition, women and men may have diverse and conflicting interests in land, and, more often than not, men's interests will take precedence over those of women. For example, men may be most interested in cash cropping, while women focus on household subsistence. If agricultural extension is directed toward cash cropping since the men are the targets, women can be excluded (USAID, 1997). Finally, because women are often excluded from decision-making processes in local communities, they may be excluded from accessing community land and the related institutions. In a study of 23 villages in the Northern Transvaal in South Africa, it was found that systems of authority actively excluded women from community meetings in almost all of the studied villages (Meer, 1997b). Indonesia's West Java provides another example, where women are often excluded from participation in 'socialisation meetings', which are the gatherings held when the ongoing land administration project is systematically implemented. These meetings are most often held in the evenings, when it is nearly impossible for women to attend (Brown, 2002).

Land-related institutions often fail to provide women and others access to information, such as the notice of the existence and scope of a right, how to register a right, how to engage in a transaction, how to fulfil obligations that sustain or support that right, and the notice and content of extension services that relate to the land. In Peru, women were frequently unaware of their legal rights to joint registration (Deere and Leon, 2001). Similarly, in regions of the Kyrgyz Republic, women generally do not know their rights regarding land, nor do they know their rights pertaining to land taxation and lease payments (Giovarelli et al., 2001).

Credit and financial institutions (although not strictly land-related) sometimes fail to provide access and use by women or other groups, and sometimes prevent women and other groups from accessing services and benefits by way of exclusionary financial requirements. For example, fees or collateral requirements can exclude some from benefiting from institutional services.

Given the disadvantaging and exclusionary effects of inadequate formal laws and biased customary laws and practices, mitigating efforts are required, which should be crafted to affect and reach not just women, but also the poor minorities, and other people on the edge of access and benefits.

Firstly, governments at all levels need to promote equity and national governments in particular need legal systems that promote and ensure gender (and other group) equality (USAID, 1997). Strengthening should not only target land programmes, but also education, health, and general rights, which might require constitutional amendments. In the case of land-related laws, joint titling to married couples and those in other consensual unions should be a focus. Legalising joint ownership and titling of land clearly establishes that rights are vested in both men and women (Deere and Leon, 2001). When women have legal ownership status, they are much more likely to interact with land institutions and to transact in land (Diana and Magdalena, 2001). However, for joint titling to be effective, it must be mandatory (Deere and Leon, 2001), since if there is an *option* to include women on titles, many households and institutions may chose not to do so. Brazil was one of the first countries to implement joint titling during agrarian reform when couples were eligible to obtain title in both of their names. However, joint titling was optional and therefore not embraced by registering institutions (Deere and Leon, 2001). Assuming mandatory joint titling, joint ownership interests should be determined when that property is registered (if they have not otherwise been formally established). Sufficient monitoring should also be undertaken by appropriate governmental institutions to ensure that this is being carried out (Giovarelli et al., 2001).

Secondly, when new laws are crafted and implemented they should build upon equitable and transferable customary laws whenever possible (USAID, 1997; Giovarelli and Akmatova, 2002). The closer statutory law conforms to customary law, the greater will be the chance is of it being accepted. In the Kyrgyz Republic, where conflicts exist between customary law and written law, customary law generally prevails in the villages because written law often fails to reflect the controlling reality of the villagers' lives. Inheritance and marriage laws should be a target for a reasonable degree of conformance (Meer, 1997a; Yngstrom, 2002).

However, a balance will be required: formal law should not blindly conform to customary law if the latter significantly disadvantages women or others. Furthermore, if the formal law conforms too closely and embodies the customary law, some of the flexibility provided by the customary law may be lost. If carefully crafted, the formal law can be used, to some extent, as an instrument to drive the customary law in the desired direction. However, formal law should not perpetuate some of the customary concepts, such as the use of shame and the loss of community respect to promote compliance. Furthermore, policy makers and legal drafters should not disregard the value and strength of these kinds of cultural norms. It has been shown that without an understanding of these fundamental norms of behaviour, formal law will not be embraced and will be irrelevant (Giovarelli and Akmatova, 2002).

Thirdly, women need knowledge of rights and obligations, and better access to dispute resolution mechanisms (Yngstrom 2002; Diana and Magdalena, 2001). Awareness campaigns (for both men and women), and programmes that provide support and assistance, are central to addressing those needs. Women can be more effectively reached through campaigns that operate during convenient hours, use female staff to do outreach and work, provide transportation services or travelling companions to women to increase mobility, publish and distribute materials in appropriate forms and places, and offer forums for women to participate in decision-making processes within institutions

Closely related to information about land are agricultural extension services, which should also be targeted to reach women. For example, incentive programmes can encourage extension agents to meet with greater numbers of women. Women extension agents should be recruited, trained, and directed to focus on working with women on increasing production and on new techniques for reducing workloads. Furthermore, women will be much less likely to be reached under extension systems that require group decision making because even if women participate in the group, they will not speak independently in front of the male participants (Eilor and Giovarelli, 2001).

Fourthly, women need greater information about and access to land-related financial institutions and services and some ideas to improve this linkage have been implemented. In Ghana, rural banks, credit unions, and NGOs are taking the lead to ensure that women have access to credit. In other places, women themselves have taken the lead in organising credit co-operatives. In Kyrgyzstan, loans with favourable terms are being extended to women so that they can create cottage agricultural processing operations (Giovarelli, 2002).

Finally, lessons need be learned from past attempts to reach women, particularly if these attempts have failed. In 1996, the government of South Africa established a sub-directorate on gender to implement mechanisms directed towards ensuring the participation of women in land reform. Unfortunately, few if any women have benefited from these efforts (Meer, 1997a). The recommendations put forth to reform and improve this initiative would serve to improve access to any land project. These include ensuring during project planning, beneficiary selection, and project appraisal that women will benefit and placing specific provisions in place for women to access financial services. Linked to these, the provision of gender training to all officials and members of organisations involved in implementing the land reform programme are essential (Meer, 1997a).

*Human resources and staff management*

Quality, efficiency, and cost-effectiveness of institutions and their products can be prompted in several ways that are related to staffing and personnel (*see* Dale and McLaughlin, 1999; Smith, 2001; Farvacque and McAuslan, 1992). Firstly, opportunities for continuing education and training for public servants are necessary to promote mobility within departments and encourage self-improvement. Secondly, in addition to basic pay and raises, a flexible pay scale that is tied to performance should be established where feasible. Such incentive-based approaches to personnel management can be used to reward productivity, cost consciousness, innovation, and effective service. Failures should be penalised with such results as decreased pay, repayment of costs, demotion, bankruptcy, marginalisation, transfer, or depending on the position, by losing a public election. Objective performance measures are central to incentive-based management, and institutional processes and services should be crafted so that such measures are a by-product. Lastly, participative management can be used to foster ownership within positions and a sense of personal responsibility.

The staffing of institutions by personnel who remain from periods during which drastically different goals or policies were being implemented, should be avoided, wherever possible since new policies and directions may necessitate new leaders and staff. As a minimum requirement, controlling bodies of institutions should be broadened to include representation that favours the new policies and directions.

*Corruption and inefficiency*

Corruption, abuse of power, patronage, nepotism, and inappropriate capture of benefits have prevented some land-related institutions from functioning as intended. For example, in Zimbabwe in the 1980s, local officials responsible for implementing a resettlement programme were accused of misbehaviour and attitudes that contributed to an overall distrust of the programme. Charges of abuse of power were ultimately levelled against the programme's director, and the overall result was a widespread perception of programme failure and a lack of necessary public support (Cliffe, 2000). In Uganda, a recent study ranked the land registry as the country's most corrupt commercial justice institution, and a recent land market study showed that perceptions of corruption in the registry prevented rights-holders from registering rights (Government of Uganda, 2001; Mwebaza and Gaynor, 2002). One of the primary ways to fight corruption is to enact legislation that guarantees the public right to information. This would allow for citizens' groups to access such information as budgets and expenditures of agencies, and to undertake public audits (Jenkins and Goetz, 1999).

Participation by regulated users, customers, and constituents can help fight corruption by giving them a better understanding of institutional decisions

(Munro-Faure, 1995; NERA, 1998). Participation can include civil society organisations, regulation or licensing by professional associations, and input from clients. Civil society organisations may participate by conducting public audits, boycotts, or taking other politically or socially charged actions (Smith, 2001). Professional associations may provide a standard of professionalism that is incompatible with corruption (Dale and McLaughlin, 1999). Direct participation by clients can provide a degree of monitoring and can also empower the clients, such as learning process approach, whereby local people can share knowledge and information with officials, to help create programmes that fit needs and institutional capacity.

Institutional simplification can work against corruption through modifying processes so that each step serves one basic function, thereby reducing the opportunities for corrupt practices (Vittal, 2001). Simplicity should not imply rigidity, and some flexibility must be allowed so that rigid rules do not lead to corruption by empowering officials who may have limited power to bend the rules (Transparency International, 2000; Dale and McLaughlin, 1999).

Organisational change may also need to take the form of eliminating dysfunctional and unnecessary departments or programmes, by rotating officers into different positions or locations, or by establishing random checks or audits of work product. In addition, a system of checks and balances between branches of government or departments that is both horizontal and vertical will help to deter corrupt practices at the intermediate level of government (Transparency International, 2000). India, for example, has established an independent government agency specifically to investigate corruption, which is designed to process accusations of corruption systematically and expediently. During the investigation, the names of those accused of corruption may be published in a public forum (Vittal, 2001). In much the same way, a declaration of any personal interests by public officials that might create a conflict of interest can reduce corruption (Transparency International, 2000).

Finally, establishing monitoring mechanisms or bodies to participate in, and review, institutional operation can promote the quality and effectiveness of institutional product. These bodies can include members from government, the grassroots, and private enterprise and such a body needs to be politically autonomous (Dale and McLaughlin, 1999). This monitoring function should be formal and routine and include ways for recommendations to become institutional practice (Farvacque and McAuslan, 1992; NERA, 1998). This can be especially useful in checking broad powers devolved to local entities (Giovarelli, 2002).

## Legal elements

Most land-related institutions are intertwined with the legal rights and obligations of citizens. Two tenets are certain to hold true when legal rights and obligations are at play: any information about the rights and obligations is central to their exercise and enforcement; and disputes will occur and will need to be resolved.

*Legal literacy*

Legal literacy, as it relates to land, refers to the knowledge and understanding of land rights and obligations, and the ability to apply those rights. Ideally, both individuals and communities would have an understanding of those land rights, which would include:

- a knowledge of land laws and the related procedures;
- an understanding of the complexity of land issues;
- some familiarity with the relationships between different laws and practices;
- a knowledge of the available transactional options;
- an understanding of the limitations of the legislation and the implementing bodies.

Land-related institutions are probably best situated to provide much of this understanding to help promote legal literacy. If the institutions do not provide it, they should work to accommodate it by striving for simplicity and transparency. A variety of tools can be used to achieve legal literacy.

An obvious central component to establishing land literacy is education, and there are a variety of considerations and related approaches (Giovarelli, 2002b). The most common attempts have been project-focused efforts to educate the public about a single activity, issue, or event. In Ukraine, for example, a current land titling project includes a significant public education component that is directed toward both a national audience and each geographic area undergoing titling (USAID, 2002).

Although project efforts can help in project implementation, their success and sustainability is dependant upon mainstream legal land literacy in the communities and society. Targets in this community-based effort beyond basic rights holders, include local community leaders, farmers, judges, NGOs, and land professionals (both public and private). The use of media selection will depend, for example, upon demographic profiles, income levels, gender, and occupations. Media will include combinations of television, radio, print, and the internet. Special tools and groups should also receive special attention. Efforts directed towards farmer and user associations (e.g., water user associations), will need to have their contents carefully crafted to reach the audience and to permit further distribution by the organisations. Efforts directed toward youth and women will need to be similarly tailored to reach the targets and be assimilated by them. Special community events, such as street theatre, can reflect popular culture and be effective in reaching the average citizen. Not surprisingly, the monitoring and review of legal literacy programmes is critical to gauge effectiveness and serve as input for modification. Monitoring and reporting programmes that focus solely on quantification should be viewed with scepticism. Numbers of meetings held, and of people reached, (e.g., through a registration sheet), numbers of television or radio spots run, numbers of items printed and supposedly circulated, can all be misleading indicators of the degree of legal literacy, and can result in a waste of resources. Instead, qualitative monitoring and appraisal efforts should be used that include, for example, interviews, before-and-after testing, and focus group debriefings.

Another important tool in promoting legal literacy and the subsequent exercise of legal rights has been the establishment of legal aid centres that can deliver both education and legal assistance, which can be used instrumentally to focus on areas or entities that have been slow to change. These centres have been successfully used to educate and promote the land rights of citizens in countries as diverse as Mexico, South Africa, Russia, and Kyrgystan (Deininger, 2001; Rolfes, 2000). Similarly, in Uganda, women have used newly-founded, legal aid centres to overcome gender-biased decisions handed down by some local dispute resolution bodies (Khadiagala, 2001).

The operation of these legal aid centres has provided useful guidelines (Bledsoe, 2001). Firstly, the staff of legal aid centres have the ability, experience, and connections in carrying out public information activities, including articles in newspapers, television coverage, informational notices/bulletins, town hall meetings, rights summaries and handbooks, and the like. Furthermore, they should be familiar with local and regional government structures and institutions and have the capacity to become involved in appropriate positions to educate officials regarding their responsibilities and the legal rights of their constituents and continue advocacy for clientele. The must also have ability and experience in legal advocacy and client counselling, and be able to demonstrate familiarity with the law and regulations (and lack thereof).

Secondly, specific legal aid centre public information and advocacy activities might include such roles as placing articles in local and regional publications, creating and posting notices and bulletins (written in appropriate languages) in local gathering places, and conducting meetings. They should also be involved in the preparation of handbooks and other similar publications, and advocacy documents, including letters to clients instructing them on their rights and recommended actions, letters of request to officials for information and administrative response, demand letters to officials, and rights pleadings. Other diverse activities might include holding meetings and consultations with clients, tribal or village chiefs, farm bosses/stock company heads, financial institution representatives, and local/regional government officials, attending advocacy events, including negotiations, settlement conferences, and court appearances.

Thirdly, a complete set of tools for the measurement of outputs is needed, and a continuous monitoring and evaluation process should be set up at an early stage. Again, care should be taken to evaluate content and efficacy, as well as quantifiable data (such as numbers of clients served).

### *Administrative dispute resolution*

The provision for some form of administrative dispute resolution for land-related controversies could be very useful depending on the circumstances within a given country, especially if land-related disputes are a burden on the formal judicial system. Any administrative dispute resolution body must have certain structural and functional aspects in order to ensure equity and efficiency, which should include: avenues for appeal to a judicial body that can check the power of the administrative body; clear, accessible, and relatively simple procedures; standardised forms; physical accessibility; a minimum of fees and costs; and the capacity to accommodate different forms of evidence (Deininger, 2001). In addition, some form of public record keeping for decisions should be established to further support transparency and accountability. Common law countries need be aware that administrative resolution of disputes will probably reduce the number of case reported that can serve as precedent. However, administrative cased could be given procedural weight if so desired (Schwarzwalder, 1999).

Just as land-related administrative institutions should work to include (and not to exclude) targeted but often-neglected groups, administrative dispute resolution forums must be equitably accessible, for example, to women, minorities, low-income populations, and under-educated persons, who must be assured of impartial outcomes (Unruh, 2001). Many of the same accessibility issues that apply to land-related institutions in general also apply to dispute resolution forums.

Legal land literacy must connect strongly with dispute resolution, so that rights holders know whether they have a contestable controversy and whether they have standing to pursue it. Again, accessibility is a function of information and the ability to use it. The appropriate tools will vary greatly from country to country and even across regions within countries. The tools will be created when policy makers and government officials acknowledge the variety of different realities and the wide range of culturally different understandings of land rights, boundaries, transactions, and uses (Deininger, 2001). For example, in some rural areas of northern Bolivia, many people relate communally to the land and understand boundaries as corresponding with the flexible relationships of communities, rather than as static lines that define absolute ownership, and thus use an informal, non-judicial dispute resolution forum to address land disputes, such as a local government official or a community authority (Goodale and Sky, 2001).

Similarly, accessibility is influenced by what is and is not deemed to constitute admissible and sufficient evidence in a land dispute. Depending on the historical development of both formal and traditional land rights in a given country, individuals may appropriately be afforded the opportunity to introduce a wide range of evidence to support their claims. In some countries, this evidence may only be in the form of oral testimony whilst on others, written records may exist that either conflict with or support the oral claims. By admitting the oral evidence, the rights of smaller holders and women may be more easily established and better protected (Deininger, 2001). In addition, through permitting oral testimony from family and community members to be admissible in both informal and formal dispute resolution institutions, together with other corroborative evidence such as physical aspects of the land, the utility and legitimacy of such institutions for small landholders and underrepresented groups will be increased (Unruh, 2001).

*Judicial dispute resolution*

Land-related administrative institutions must also be supported by judicial dispute resolution mechanisms, both as avenues for appeal and for venues of first consideration. In some developing and transitional countries, land disputes make up the bulk of the claims within the court systems and, therefore, a functioning judicial system is a correspondingly important asset. In Cambodia, for example, it has been reported that more than 50 per cent of court cases are related to land conflicts, whilst in Indonesia, the number is closer to 65 per cent (Deininger, 2001). A cycle of lost productivity can be created as insecure property rights lead to disputes, which are then taken to an overburdened judicial system for costly and time-consuming resolution and, in the meantime the land that is in dispute is unproductive. Some or all of the income or sustenance that might have been reaped from the land is lost.

The countries that are developing land administration capabilities may or may not have viable judicial systems in place since some lack the fundamentals of a workable judicial system and the basic frameworks that support and reflect the rule of law (Bledsoe, 2000). Many countries making the transition from centrally planned economies to market economies are left only with remnants of an essentially inoperable socialist legal structure (Prosterman and Hanstad, 1999). Others may have systems that operate reasonably well to resolve some form of criminal, contract, or civil disputes, but are not effective in resolving land disputes. The reasons for the failure include: the novelty and specialisation of land law and land disputes because of correspondingly new land rights and the formalisation of these rights; and a failure, given limited resources, to prioritise or accommodate the adjudication of land disputes.

In Uganda, for example, the progressive 1998 Land Act provides for a sophisticated system of land courts distributed throughout the country and operated by local officials. However, because of funding and staffing limitations, very few of these courts had been established by late 2001. The judicial land dispute mechanism is now being modified to provide for a more realistic structure (Republic of Uganda, 2001b). Under any circumstances, working towards reforming the judicial system and its institutions is a long and involved process for any country (Messick, 1999).

In some cases, capital investments in technology or facilities can be less valuable than focusing on consensus building to support the establishment of the rule of law. Such consensus building or structural reforms may include developing legal aid clinics, ensuring that judicial appointments are merit-based, and encouraging a public discussion about the law and its enforcement (Messick, 1999). Other reforms can include establishing the judicial branch as an independent institution capable of holding the other branches of government accountable, making the process of filing a suit more efficient, decreasing caseloads, increasing access by establishing some alternative dispute resolution forums, and making the bench and bar more professional (Messick, 1999).

## References


Admassie, Y., 2000. *Twenty Years to Nowhere: Property Rights, Land Management, and Conservation in Ethiopia*, The Red Sea Press, Inc., New Jersey.

Agarwal, A. and Ostrom, E., 2001. Collective action, property rights, and decentralisation in resource use in India and Nepal, *Politics and Society*, **29**(4), 485-514.

Alexander, J., 2000. The Enduring appeal of 'technical development' in Zimbabwe's agrarian reform. In: Bowyer-Bower, T. and Stoneman, C. (eds.), *Land Reform in Zimbabwe: Constraints and Prospects,* Ashgate Publishing Ltd, Aldershot.

Australian Agency for International Development, 2000. *Improving Access to Land and Enhancing the Security of Land Rights: A Review of Land Titling and Land Administration Projects*, Quality Assurance Series, No. 20, Canberra.

Barnes, G., Stanfield, D. and Barthel, K., 2000. Land registration modernisation in developing economies: a discussion of the main problems in Central/Eastern Europe, Latin America, and the Caribbean', *Urban and Regional Information Systems Association (URISA) Journal*, **12**(4), 27-36.

Baaklini, A., 2001. Administration in developing countries and the democratic challenge, *Perspectives on Global Development and Technology*, 17(2), 57-69.

Barnes, G. and Eckl, M., 1996. *A GPS Methodology for Cadastral Surveying in Albania: Phase II*, Final Report submitted to Land Tenure Center, University of Wisconsin, Madison.

Barry, M. and Fourie, C., 2002. Analysing cadastral systems in uncertain situations: a conceptual framework based on soft systems theory, *International Journal of Geographical Information Science,* **16**(1), 23-40.

Binkerhoff, D. and Crosby, B., 2002. *Managing Policy Reform: Concepts and Tools for Decision-Makers in Developing and Transitioning Countries*, Kumarian Press, Connecticut.

Bledsoe, D., 2001. *Unpublished Memorandum: Legal Aid Center Staffing and Operations*, Rural Development Institute, Seattle.

Bledsoe, D., 2001. *Policy, the Rule of Law, and Rural Land Reform in China*, RDI Report on Foreign Aid and Development No. 101, Rural Development Institute, Seattle.

Brown, J., 2002. *Registration and Women's Land Rights on Java: Why so Many Married Couples Register Marital Property in the Name of One Spouse and What has been the Impact on Women's Land Rights* (draft on file at Rural Development Institute).

Brown, J., 2000. Land reform vs. customary law: what about women? In: Bowyer-Bower, T. and Stoneman, T. (eds.), *Land Reform in Zimbabwe: Constraints and Prospects*, Ashgate Publishing Ltd, Aldershot.

Bruce, J., 1986. *Land Tenure Issues in Project Design and Strategies for Agricultural Development in Sub-Saharan Africa*, Land Tenure Center, University of Madison-Wisconsin.

Bruce, J., 1999. *Legal Bases for the Management of Forest Resources as Common Property*, Food and Agriculture Association of the United Nations, Rome.

Cliffe, L., 2000. The politics of land reform in Zimbabwe. In: Bowyer-Bower, T. and Stoneman, C. (eds.), *Land Reform in Zimbabwe: Constraints and Prospects*, Ashgate Publishing Ltd, Aldershot.

Cooper, P., Brady, L., Hidalgo-Hardeman, O., Hyde, A., Naff, K., Ott, J. and White, H., 1998. *Public Administration for the Twenty-First Century*, Harcourt Brace, Fort Worth.

Dale, P. and McLaughlin, J., 1999. *Land Administration*, Oxford University Press, Oxford.

Deere, C. and Leon, M., 2001. *Empowering Women: Land and Property Rights in Latin America*, University of Pittsburgh Press, Pittsburgh.

Deininger, K. (ed.), 2001. *Land Policy and Administration: Lessons Learned and New Challenges for the Bank's Development Agenda*, World Bank, Washington, D.C.

Deininger, K. and May, J., 2000. *Is there Scope for Growth and Equity: The Case of Land Reform in South Africa* (CSDS Working Paper No. 29), School of Development Studies, Durban.

Du Guerny, J. and Topouzis, D., 1996. *Gender, Land and Fertility – Women's Access to Land and Security of Tenure*, FAO, http://fao.org/sd/wpdirect/wpan0001.htm (visited 8/21/02).

Eilor, E. and Giovarelli, R., 2001. *Land Sector Analysis: Gender Component*, Government of Uganda.

Enemark, S. and Sevatdal, H., 1999. Cadastres, land information systems and planning – is decentralisation a significant key to sustainable development? *UN-FIG Conference on Land Tenure and Cadastral Infrastructures for Sustainable Development*, Melbourne, Australia.

Ericsson, A., 1999. Women's access to land, *UN-FIG Conference on Land Tenure and Cadastral Infrastructures for Sustainable Development*, Melbourne, Australia.

FAO, 1997. *Women and Land Tenure*, FAO Land Tenure Series, UN-Food and Agriculture Organisation, Rome.

FAO, 2002. Guidelines on Rural Property Tax Systems in Central and Eastern Europe, FAO Land Tenure Series, UN-Food and Agriculture Organisation, Rome.

Farvacque, C. and McAuslan, P., 1992. *Reforming Urban Land Policies and Institutions in Developing Countries*, World Bank Urban Management Programme Policy Paper 91, World Bank, Washington, D.C.

Firman-Sellers, K. and Sellers, P., 1999. Expected failures and unexpected successes of land titling in Africa, *World Development*, 27(7), UN-Food and Agriculture Organisation, Rome.

Fox, W., 1994. *Strategic Options for Urban Infrastructure Management*, World Bank Urban Management Programme Policy Paper 52, World Bank, Washington, D.C.

Gaynor, R. and Bledsoe, D., 2000. *Evaluation of the Albania Land Market Project*, U.S. Agency for International Development.

Gerber, L., 2002. *Land Adjudication in Andhra Pradesh*, Unpublished memorandum on file with Rural Development Institute.

Giovarelli, R., 2002a. Interview, based on fieldwork in the Kyrgyz Republic, July 29, 2002.

Giovarelli, R., 2002b. *Land Literacy Strategies: Some Possibilities* (Draft), UN-Food and Agriculture Organization, Rome.

Giovarelli, R., Aidarbekova, C., Duncan, J., Rasmussen, K. and Tabyshalieva, A., 2001. *Women's Rights to Land in the Kyrgyz Republic*, World Bank, Washington, D.C.

Giovarelli, R. and Akmatova, C., 2002. *Local Institutions that Enforce Customary Law in the Kyrgyz Republic and their Impact on Women's Rights*, World Bank, Washington, D.C.

Girishankar, N. and De Silva, M., 1998. *Strategic Management for Government Agencies: An Institutional Approach for Developing and Transition Economies*, World Bank Discussion Paper No. 386, World Bank, Washington, D.C.

Gong, T., 2002. Dangerous collusion: corruption as a collective venture in contemporary China, *Communist and Post-Communist Studies*, **35**, 85-103.

Goodale, M. and Sky, P., 2001. A comparative study of land tenure, property boundaries, and dispute resolution: case studies from Bolivia and Norway, *Journal of Rural Studies*, **17**, 183-200.

Gray, L., 2002. Environmental policy, land rights, and conflict: rethinking community natural resource management programmes in Burkina Faso, *Environment and Planning D, Society and Space*, **20**, 167-182.

Government of the Kyrgyz Republic, 1999. *Land Code of the Kyrgyz Republic.*

Hayami, Y., 1994. Strategies for the reform of land property relations. In: Barker, R. (ed.), *Agricultural Policy Analysis for Transition to a Market-Oriented Economy in Vietnam*, FAO Economic and Social Development Paper 123, UN-Food and Agriculture Organisation, Rome.

Hennessy, P., 1998. Hurd calls for a new EU treaty to protect the Nation State, *London Evening Standard*, June 17, 1998, 19.

Hentic, I. and Bernier, G., 1999. Rationalisation, decentralisation and participation in the public sector management of developing countries, *International Review of Administrative Sciences*, **65**(2), 197-210.

Ho, P., 2001. Who owns China's land? Policies, property rights and deliberate institutional ambiguity, *The China Quarterly*, **166**, 394-421.

Jacobs, S., 2000. The effects of land Reform on gender relations in Zimbabwe. In: Bowyer-Bower, T. and Stoneman, C. (eds.), *Land Reform in Zimbabwe: Constraints and Prospects*, Ashgate Publishing Ltd, Aldershot.

Jenkins, R. and Goetz, A.-M., 1999. Constraints on civil society's capacity to curb corruption: lessons from the India experience, *IDS Bulletin*, **30**(4), 39-49.

Jreisat, J., (ed.), 2002. *Governance and Developing Countries*, International Studies in Sociology and Social Anthropology Series, Brill, The Netherlands.

Khadiagala, L., 2001. The failure of popular justice in Uganda: local councils and women's property rights, *Development and Change*, **32**, 55-76.

Kigula, J., 1993. Land Disputes in Uganda: An Overview of the Types of Land Disputes and the Dispute Settlement Fora', *Access to Land and Other Natural Resources in Uganda: Research and Policy Development Project*, Land Tenure Center Research Paper 3, Land Tenure Centre, University of Madison-Wisconsin.

Larson, A., 2002. Natural resources and decentralisation in Nicaragua: are local Governments up to the job? *World Development*, **30**(1), 17-31.

Mackinnon, J. and Reinikka, R., 2000. *Lessons From Uganda on Strategies to Fight Poverty*, World Bank, Washington, D.C.

MacNeill, T., Ivan, F. and McGrath, G,, 1998. Implementing real property systems in the former Soviet Union, *Land Use Policy*, **15**(1), 45-53.

Meer, S., 1997a. Gender and land rights: the struggle over resources in post-apartheid South Africa, *IDS Bulletin*, **28**(3), 133-144.

Meer, S., (ed.), 1997b. *Women, Land and Authority: Perspectives from South Africa (OXFAM)*, Creda Press, Cape Town.

Messick, R., 1999. Judicial reform and economic development: a survey of the issues, *The World Bank Research Observer*, **14**(1), 117-136.

Morshed, M., 1997. *Bureaucratic Response to Administrative Decentralisation: A Study of Bangladesh Civil Service*, The University Press Limited, Dhaka, Bangladesh.

Munro-Faure, P., 1995. Supporting markets in the agricultural land of transitional economies. In: Wunderlich, G. (ed), *Agricultural Land Ownership in Transitional Economies*, University Press of America.

Munyakho, D., 1994. Kenyan women press for land rights, *African Farmer*, **8**, April.

Muhereza F. and Bledsoe, D., 2002. *Land Sector Analysis: Common Property Resources Component*, Government of Uganda.

Mwebaza, R. and Gaynor, R., 2002. *Land Sector Analysis: Land Market Component*, Government of Uganda.

Mwita, J., 2000. Performance management model: a systems-based approach to public service quality, *The International Journal of Public Sector Management*, **13**(1), 19-37.

National Economic Research Associates, 1998. *Governance and Regulatory Regimes for Private Sector Infrastructure Development*, **21**, Asian Development Bank.

Pearce, J., (ed.), 2000. *Development, NGOs, and Civil Society*, Oxfam, Oxford, U.K.

Prosterman, R. and Riedinger, J., 1987. *Land Reform and Democratic Development*, John Hopkins University Press, Baltimore.

Prosterman, R. and Hanstad, T. (eds.), 1999. *Legal Impediments to Effective Rural Land Relations in Eastern Europe and Central Asia*, World Bank Technical Paper No. 436, World Bank, Washington, D.C.

Rattanabirabongse, V., Eddinton, A. and Nettle, K., 1998. The Thailand land titling project – thirteen years of experience, *Land Use Policy*, **15**(1), 3-23.

Republic of Uganda, 2001a. *Uganda Commercial Justice Baseline Study*, Law and Development Partnership Limited and Steadman Research Services.

Republic of Uganda, 2001b, *Land Sector Strategic Plan*, Final Draft, November 30, 2001.

Riddell, J., 1998. Private/public partnerships in land administration in Black Sea countries (Part 1), *Sustainable Development Dimensions*, SD Conference, Bertinoro Italy.

Robertson, W., 2002. Anticipating the further development of cadastral systems, *Computers, Environment, and Urban Systems*, **26**, 383-402.

Rodrik, D., 2000. What drives public employment in developing countries, *Review of Development Economics*, **4**(3), 229-243.

Rolfes, L., 2000. *Unpublished memorandum: Samara and Vladimir legal aid centre status report*, Rural Development Institute, Seattle.

Roy, K. and Tisdell, C., 1998. Good governance in sustainable development: the impact of institutions, *International Journal of Social Economics*, **25**, 1310-1325.

Rubin, P., 1999. Growing a legal system in post-communist economies, *Cornell International Law Journal*, **27**(1), 1-47.

Schwarzwalder, B., 1999a. Land registration. In: Prosterman, R. and Hanstad, T., (eds.), *Legal Impediments to Effective Rural Land Relations in Eastern Europe and Central Asia*, 165-192, World Bank Technical Paper No. 436, World Bank, Washington, D.C.

Schwarzwalder, B., 1999b. Land-related judicial institutions. In: Prosterman, R. and Hanstad, T. (eds.), *Legal Impediments to Effective Rural Land Relations in Eastern Europe and Central Asia*, 279-290, World Bank Technical Paper No. 436, World Bank, Washington, D.C.

Smith, L., 2001. *Reform and Decentralisation of Agricultural Services: A Policy Framework*, FAO Agricultural Policy and Economic Development Series 7, UN-Food and Agriculture Organisation, Rome.

Smoke, P., 1994. *Local Government Finance in Developing Countries: The Case of Kenya*, Oxford University Press, Nairobi.

South African Government Information, 2002. *Land White Paper: Institutional Arrangements*, http://www.polity.org.za/govdocs/white_papers/landwp06.html (visited 7/29/2002).

Stanfield, D., 1996. Creation of land markets in transition countries: implications for the institutions of land administration, International Conference on Land Tenure and Administration, Orlando, Florida.

Sundar, N., 2001. Is devolution democratisation? *World Development*, **29**(12), 2007-2023.

Ting, L. and Williamson, I., 1999. Land administration and cadastral trends: the impact of the changing humankind-land relationship and major global drivers, *UN-FIG Conference on Land Tenure and Cadastral Infrastructures for Sustainable Development*, Melbourne, Australia.

Toulmin, C. and Quan, J., 2000. *Evolving Land Rights, Policy, and Tenure in Africa*, Department for International Development, London.

Transparency International, 2000. Chapter 3: developing responses, *TI Source Book 2000*, www.transparency.org/sourcebook/03.html (6/28/02).

Turner, M., (ed.), 1999. *Central-local Relations in Asia-Pacific*, Macmillan Press, London.

Turner, M. and Hulme, D., 1997. *Governance, Administration and Development: Making the State Work*, Kumarian Press, West Hartford, Connecticut.

UNCHS, 1999. *Women's Rights to Land, Housing and Property in Post-Conflict Situations and During Reconstruction.* United Nations Centre for Human Settlements, http://www.unchs.org/campaigns/tenure/women.asp (8/21/02).

UNECE, 1996. *Land Administrative Guidelines: With Special Reference to Countries in Transition*, United Nations Economic Commission for Europe, New York.

UNECEAP, 1995. *Municipal Land Management in Asia: A Comparative Study*, United Nations Economic and Social Commission for Asia and the Pacific, New York.

UNFPA, 2002. *Women as Food Producers*, United Nations Population Programme, New York, http://www.unfpa.org/modules/intercentre/food/womenas.htm (7/31/2002).

Unruh, J., 2001. Post war land dispute resolution: land tenure and the peace process in Mozambique, *International Journal on World Peace*, **18**(3), 3-30.

USAID, 1997. *Information Bulletin – The Right to Own Land – A Fundamental Principle of Development*, USAID, Washington, D.C.

USAID, 2000. *Decentralisation and Democratic Local Governance Programming Handbook*, http://www.usaid.gov/democracy/pdfs/pnach300.pdf (8/21/02).

Van der Molen, P. and Osterberg, T., 1999. Land tenure and land administration for social and economic development in (Western) Europe, *UN-FIG Conference on Land Tenure and Cadastral Infrastructures for Sustainable Development*, Melbourne, Australia.

Vittal, N., 2001. Corruption and the state, *Harvard International Review*, **23**(3), 20.

Wallis, J. and Dollery, B., 2001. Government failure, social capital and the appropriateness of the New Zealand model for public sector reform in developing countries, *World Development*, **29**(2), 245-263.

Weinberger, K., 2001. Women's participation in local organisations: conditions and constraints, *World Development*, **29**(8), 1391-1404.

Williamson, I., 2001. Land administration "best practice" providing the infrastructure for land policy implementation, *Land Use Policy*, **18**, 297-307.

World Bank, 2000a. *Project Appraisal Document on Proposed Credit to the Kyrgyz Republic for a Land and Real Estate Registration Project*, http://www-wds.worldbank.org/servlet/WDSContentServer/WDSP/IB/2000/06/10/000094946_0005 2005302366/Rendered/PDF/multi_page.pdf (visited 7/29/02).

World Bank, 2000b. *Reforming Public Institutions and Strengthening Governance: A World Bank Strategy (November 2000)*, World Bank, Washington, D.C.

World Bank, 2001. Decentralisation and governance: does decentralisation improve public service delivery? *PREM Notes*, **55**, June 2001,

World Bank, 2002a. *A User's Guide to Poverty and Social Impact Analysis, Draft for Comment*, Poverty Reduction Group and Social Development Department, World Bank, Washington, D.C.

World Bank, 2002b. Decentralisation,
http://www1.worldbank.org/publicsector/decentralisation/Different.htm
http://www1.worldbank.org/publicsector/decentralisation/Rationale.htm (8/21/02).

Yngstrom, I., 2002. Women, wives and land rights in Africa: situating gender beyond the household in the debate over land policy and changing tenure systems, *Oxford Development Studies*, 30(1), 21-40.

Chapter 8

# Managing Informal Settlements in Botswana

R. W. Dixon-Gough and Otladisane Molobeng

## Background

All countries throughout the world have experienced mass migrations from rural to urban areas, with virtually half the total population now living in urban areas (GUO, 2003). This proportion is likely to grow in the immediate future and will lead to the urbanisation of social, economic, and environmental issues such as poverty and pollution (Wallis and Mumtaz, 2004).

Urban areas have different activities, services, and population concentrated in relatively small areas, and are often geographically and spatially constrained. Land is, therefore, an important and scarce commodity, and its use has to be regulated through physical planning. This entails determining what activity should take place where, when, and how. Thus, the physical planning of an urban area determines the nature of ownership, its area, and its location. Land ownership is also regulated through legislation and arrangement, such as its use through physical planning and enforcement laws. If an urban parcel of land or lot is acquired and developed without the regulatory laws of the city or town, it is termed informal. If it is for housing purposes, then it gives rise to an informal settlement.

Informal settlements may be referred to in different ways in different regions of the world. They are often internationally known as squatter settlements, shantytowns, or slums. However, they are specifically called, for example, *majengos* in Kenya (Majale, 1998), *favelas* in Brazil (De Sampaio, 1994); and *gecekondus* (Malusardi and Occchipinti, 2003) in Turkey. They are typified as dense settlements comprising communities housed in self-constructed shelters, under conditions of informal or traditional land tenure. They are defined by the UN Habitat Programme of 1995 as:

- residential areas where a group of housing units has been constructed on land to which the occupant has no legal claim or which they occupy illegally; and
- unplanned settlements and areas where housing is not in compliance with current planning and building regulations (unauthorised housing).

They are common features of developing countries and are typically the product of an urgent need for shelter by the urban poor. As such they are characterised by a dense proliferation of small, makeshift shelters built from diverse materials, degradation of the local ecosystem, and by severe social problems. Informal settlements tend to occur when the current land administration and planning fails to address the needs of the whole community. These areas are characterised by rapid, unstructured and unplanned development. On a global scale, informal settlements are a significant problem housing the world's disadvantaged.

The problems that this will generate can be witnessed in the growth of informal settlements in which (for most cases) there is limited access to basic services such as drinking water and basic sanitation. This is of particular importance in the case of many large cities where a large proportion of the population, who provide the cities with a resource of cheap labour, live in slums or informal settlements (Wallis and Mumtaz, 2004). The growth of such settlements has become a major issue in many countries and some have provided programmes to regularise and upgrade the settlements. Informal settlements vary in their characteristics and size. Some have populations of over 100,000, whilst others are very small and occupy infill sites (Pugh, 2000).

India provides a complete contrast to many African countries. On one hand it is a country of extreme technical ability (witness its earth resources satellite programme), yet on the other, it is a country of extreme poverty. Most of this poverty is encountered in urban areas, the population of which have been expanding since 1950 at the rate of about 7 to 8 million per year. At present, approximately 93 per cent of the 290 million urban workforce live in the informal sector (Planning Commission, 2003). Most of this workforce is self-employed, casual, contract or migrant and without them, the economy of India would cease to function, yet the problems of informal and slum settlements are largely ignored. One alternative would be the provision of low cost housing, but planners fail to take into consideration the basic economic realities of this provision, whilst urban land managers provide little in the way of drinking water and sanitation since these basic amenities cannot be provided unless the occupants have legal title to the land (Planning Commission, 2003).

In the case of India, and virtually every other country in which informal settlements exist, the urban migrants who are driven to the cities to escape extreme rural poverty, conflict, extreme climatic conditions, or simply unemployment, have no legal access to housing or employment. There is also an additional problem in India since 'homelessness' is a crime and vagrants, pavement dwellers, and 'illegal' squatters are routinely moved by the police for violating penal statutes and their dwellings and processions destroyed (Planning Commission, 2003).

The situation in Botswana is no exception as is evident from the number of informal settlements that have developed around the peripheries of the formal urban areas. The government and the local authorities had the task in finding ways of controlling and formalising them through plot rationalisation and resettlement, around the cities of Francistown and Gaborone, and as an extreme case, eviction (Molobeng, 2004).

Although informal settlements can be regarded as nuisance by the authorities, it must be accepted that this a part of the urban growth, which should be catered for by having enough serviced land parcels or housing sites available at affordable rentals to the lowly paid. In Botswana, the major problem with the informal settlements is due to who should be responsible for them since urban, rural and freehold land have different management and administrative structures and it is not uncommon for a single informal settlement to evolve on two, or all three land types.

There has been rapid urbanisation in Botswana. At independence in 1966, the urban areas accounted for 4 per cent of the population; in 1971 this had increased to 10 per cent and by 2001, 52 per cent of the population were urban dwellers (CSO, 2003) (Figure 8.1). Most of this urbanization is centred on the eastern side of the country, which is the most fertile part and highly developed. These figures include urbanized villages which are located in rural or tribal areas.

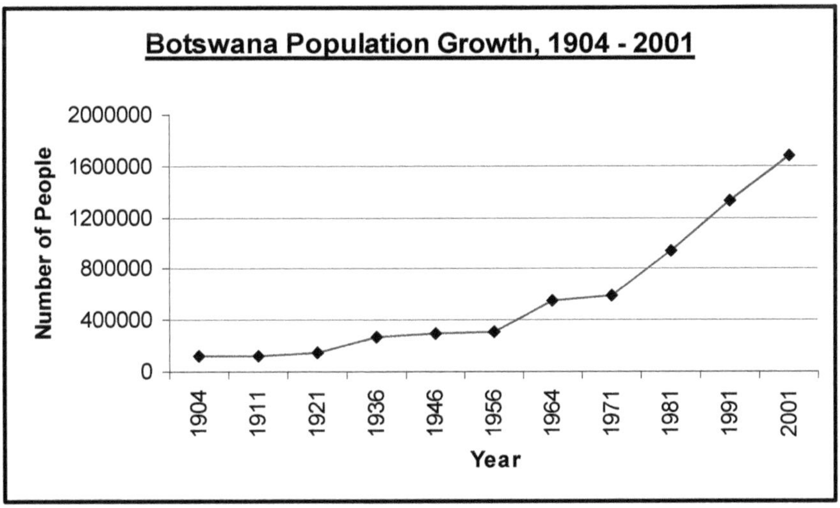


**Figure 8.1 Botswana population growth, 1904-2001** (from Molobeng, 2004)

The population of Francistown shows a population growth from 8,521 in 1964, to 83,023 in 2001 with a sudden increase in population between 1964 and 1971, which later slowed down during the 1970s (Figure 8.2). This might have been due to drought, which compels people to migrate to towns since they are unable to make a livelihood from agriculture in the rural areas. Francistown, as the first town in Botswana, has had a long history of informal or squatter settlements due to its attractiveness to migrants from neighbouring areas. Initially, it had no effective planning control within the township boundary as any land that was found undeveloped was deemed suitable by the newly arrived migrants, as a potential site for erecting houses.

Historically, most of the land for residential purposes within the township was dominated by informal settlements, which were populated by Africans. The residential development was racially segregated with Europeans in the town centre - Central, the Coloureds (mixed race people) on the eastern periphery – Satellite; and the African on the western side of the Tati River – these being Bluetown, Madzibalori, Riverside, Kgaphamadi and Somerset, and in Monarch on the northern side (Figure 8.3).

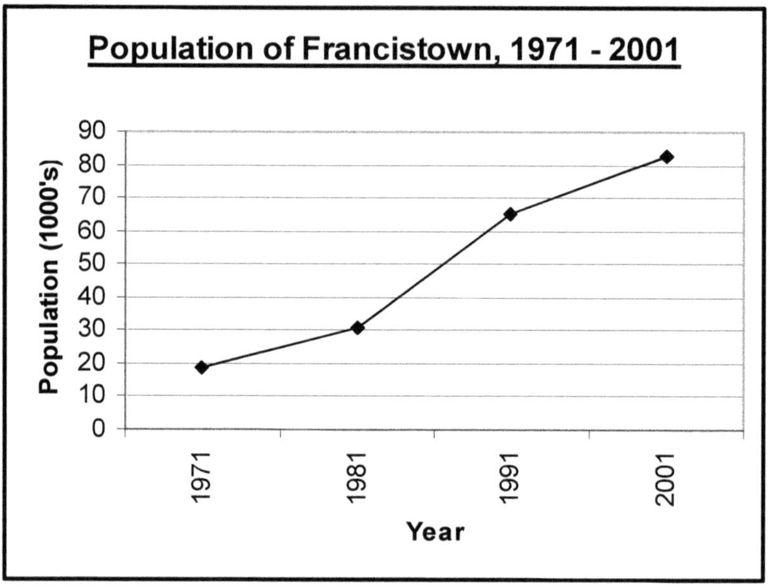


**Figure 8.2 Francistown population growth, 1971-2001** (from Molobeng, 2004)

Prior to the upgrading, the residents of the settlements understood that without a certificate from the council, they could not claim ownership of the land they were occupying and, therefore, could not invest by erecting permanent structures.

One of the problems experienced universally with informal settlements is that when they extend the area s of existing, formal settlements they create infrastructural problems such as the provision of drinking water, sanitation, education, health care, and safety problems. As the informal settlements expand, more pressure is applied to finding solutions to those problems, and one of the key management tools is the provision of accurate spatial data from which the problems can be visualised and strategies developed (Newman and Venter, 1999).

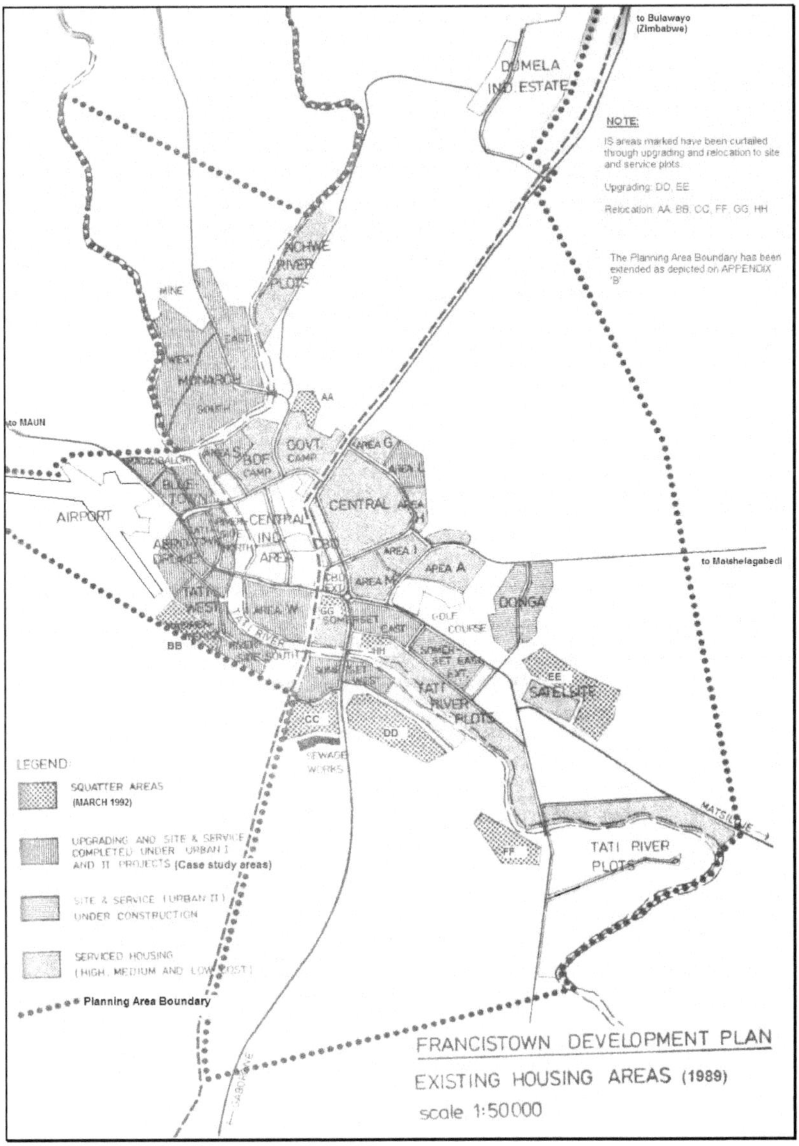


**Figure 8.3 The location of the settlements within Francistown**
*Source*: Department of Town and Regional Planning.

## Land tenure and land administration in Botswana

Botswana, like the other countries in southern Africa, inherited a dual system of statutory and customary tenure at independence. The land administration, coupled with political stability has evolved since then in such a way that has avoided many of the land-related problems that have been experienced in neighbouring countries. When Botswana attained its independence in 1966 when it was one of the poorest countries in the world, its land administration was considered to be its only valuable asset other than cattle rearing.

In 1966, Botswana had a relatively modest sized population, and no impairing legacy of colonial settlement. The success of its independence was assured by the discovery, in 1967, of diamonds which contribute to about 75 per cent of national export revenues, 33 per cent of the GDP, and 50 per cent of all direct revenue flowing into the national Treasury. It should be noted here that the Mines and Minerals Act of 1999 govern the exploration activities by setting out guidelines for prospecting and mining licences, environmental protection measures, royalties and mining taxation (GOB, 2003).

### *Secure tenure*

Even though the granting of secure tenure does not resolve the problems of land for housing or residential purposes, it does create conditions that enable people to exercise their rights and to discharge their obligations as urban inhabitants. Security of tenure is agreement involving an individual or group to land and residential property, which is governed and regulated by a legal and administrative framework. The security is derived from the right to access to and usage of the land or property being guaranteed by a set of rules. This can be affected in a variety of ways, depending on legal frameworks, social norms, cultural values, and individual preference.

A person is deemed to have secure tenure if he or she is protected from involuntary removal from their land or property, except in certain circumstances. These circumstances should be through known and agreed legal procedures, which have to be objective, equally applicable, contestable, and independent.

### *Land tenure and land administration*

Land tenure systems are one of the many tools used to regulate society. Many species of animals observe very definite rights of tenure either in the individual or in the group. Should these rights be threatened, for example through overpopulation or natural disasters, the individual or group will react fiercely to defend them. The complexities and variations of human society demand equally complex and varying rules of rights of tenure and rights to use land, which is probably the most basic of human resources (Dickson, 1982). It is an old Tswana ideal that:

> the land is like the air, in sufficient quantity, free for everyone to use as they wish.

That this ideal is no longer applicable in many parts of the world is the basic reason for the increased use of complex rules to control rights in land and the use of land.

As an example, the conversion of some communal land to urban or individual holdings is happening in several African countries, including Botswana. These issues are of major proportions, but other less violent changes in society can affect the tenure rules adopted by the government to suit community views and needs. Economic needs and changing living patterns are examples of this. For example, in Botswana, the Certificate of Rights form of tenure was created both to meet an economic need and to face the changing living pattern of large numbers of the population who have moved to an urban way of life.

There are three important administrative procedures and legislation governing tenure in Botswana. These are:

- tribal or customary or communal land which constitutes of about 71 per cent of the country's land surface;
- state land which occupies 23 per cent of the land surface; and
- freehold which is 6 per cent of the country's land surface.

*State land* This is directly administered by the central government and includes Game Reserves, National Parks, Cities, Towns, Forest Reserves, and Wildlife Management areas. Prior to independence, most of this land was under freehold but, due to the demands of land for the expansion of towns after independence, this was curtailed and the ownership of this land is now under state control. There are two main types of tenure systems in urban types:

- *Fixed Period State Grant (FPSG)*, which is a lease with the total charges paid at the commencement of the lease rather than periodically. The charges vary from location to location and according usage and area. It covers all land in urban areas, including residential, commercial, industrial, and multi-purpose parcels, and is registered under the Deeds Registry Act. At the end of the lease period, which is usual 50 or 99 years, it reverts to the state although there is a provision for renewal. All improvement, which has been made to the parcel, becomes the property of the state without any compensation if there is no renewal. The lease is transferable only on condition that building covenant has been met and what is transferred is only the grant period left. As present, all land in urban areas are surveyed prior allocation.
- *Certificate of Rights (COR)* were initiated in the 1970s to provide tenure security for the urban informal settlements and plots allocated through the council's Self Help Housing Agencies. This was a means of minimising the costs of title registration and it can be inherited, ceded, assigned, and transferred with the council's consent. It can be converted to FPSG after being surveyed though, nowadays, all plots in urban areas are surveyed prior to allocation and a payment made by the allottee to obtain a FPSG title. This is the title, which was conferred on Francistown's informal settlements

following the upgrading exercise. The issuing of a COR has been discontinued and replaced with FPSG as all plots in urban areas are now surveyed.

*Freehold land*

Freehold is the highest form of estate in land obtainable by an individual. In theory, it is complete ownership, but, in practice, today it is related to legislation (such as town and country planning regulations) in that the freedom of the freeholder to do what he/she likes with the land is limited. Nevertheless, the freeholder normally has the right to dispose of the land, pledge it as security, and attend to the long-term financial development of the land entirely as his/her own responsibility and at his/her own risk.

Theoretically, freehold is the most creditworthy of tenures. It is, however, subject to current political tendencies and pressures. For example, a freehold issued by a previous administration and contrary to the political opinions of a current government that has promised land reform and new land legislation, would probably be viewed as less secure in the short term than a lease issued by a current government. Most of the freehold land in Botswana is found in Francistown and the surrounding Tati farms; the Tuli block farms along the eastern border with South Africa; some parts of Gaborone and Lobatse; and the Ghanzi farms. Most belong to private companies and individuals though the Government has been purchasing some as to alleviate the shortage of land especially in the north-east Administrative District.

*Tribal land (communal or customary land)*

Before the inception of Land Boards (LBs), the allocation of land in different tribal territories was under the control of Chiefs and this was a considerable task for them since regional social and economic development could not be dictated by a single person. This lead to the formation of the LBs, which would be empowered to administer tribal land according to the Tribal Land Act (TLA) of 1968. Their core functions are:

- land administration, essentially involving the administration of land allocation process by undertaking such tasks as the issuing of certificates and leases, the arbitration of disputes, and recording minutes of various meetings;
- surveying and registration involving the provision of professional and technical services advice on land use planning, surveying, land and plot demarcation, assessments and valuations, bore hole investigations, and beacon identification.

With rapid urbanisation of Botswana, more land for the urban areas will be needed and it is likely that this must be made available from tribal and freehold land. There is, therefore, a need to have a common practice through which there will be easy acquisition of land, that will require a relaxation of planning and land administration procedures. In Botswana, this has been recognized and has evolved into the National Settlement Policy.

Thus the government has to make concerted effort in the improving of standard of living for those living in the tribal areas in terms of infrastructure and services which are not in par with those in urban centres. Through this, the rate of migration to urban centres will be stemmed or reduced and ultimately reduce the proliferation of informal settlements

*Informal settlements in Francistown*

Francistown, as the first town in Botswana, has had a long history of informal settlements. Initially, there was no effective planning control within the township as any land found undeveloped was deemed suitable by the newly arrived migrants, as a potential site for building shelters and houses. Most of the residential land within the township was dominated by informal settlements. The residential development was racially segregated with Europeans in the town centre (Central), the Coloureds (mixed race people) on the eastern periphery (Satellite), and the Africans on the western side of the Tati River at Bluetown, Madzibalori, Riverside, Kgaphamadi, and Somerset, and in Monarch on the northern side (Figure 8.3).

These settlements lacked virtually all facilities and the houses were constructed of mud walls with thatch or plastic roofs. The physical infrastructure then available was characterised by:

- water supplies below minimum health and urban standards;
- inadequate or non-existence sanitation with toilets in the nearby bushes or pit latrines, which were liable to collapse at any time due to their shallow depths;
- infrequent refuse collection, where available;
- non-existence of roads and storm water drainage;
- lack of street lighting for security at night; and an
- absence of facilities such as schools and clinics.

Before the upgrading, the residents of the settlements were fully aware that without a certificate from the council, they could not claim ownership of land they were occupying and therefore could not invest in their property by erecting permanent structures.

*Upgraded areas*

The locations which were upgraded were Bluetown, Madzibalori, Monarch, Riverside North and South, Somerset East and West, and Tati West. These areas provided about 2,050 plots for which the occupants were issued with COR (Figure

8.4) with the agreement that service levy charges would have to be paid monthly. This charge was for the costs of services provided by the council.

Upgrading was chosen because of many mitigating circumstances, the major one being the low availability of plots and service areas where the residents of the informal settlements could be relocated. The strong community ties, which had been forged over time, made the residents reluctant to be relocated. Since the upgrading, coupled with voluntary reallocation, would be carried out with minimal disruption the council, the government, and the donor (the World Bank) agreed to this method as the most viable one for most of the locations.

The method used for the demarcation of plots was through the use of aerial photographs, printed at a scale of 1:1,000 as to show most ground details. The aerial photographs were taken after the density of the plots had been reduced to the 320 $m^2$ minimum requirement. The residents were subsequently compelled to fence the plots or put stones along their plot boundaries that would show up in the aerial photographs. The images were useful as minimal fieldwork was necessary on those properties where boundary was not defined. This technique of mapping was undertaken for all the areas except Monarch, which was used as a reception area and as a conventional site and service area.

The reception area philosophy was the creation of areas for the very poor from the informal settlements who could not afford to pay for services provided in the upgraded areas. This was necessary due to the Government's policy of non-subsidisation, which states that whenever there is a development programme in urban areas, it should be economically affordable by the beneficiaries and that there would not be any Government subside for housing in urban areas (GOB, 1982).

The layout for Monarch followed to rearrange the plots as regularly as possible and to make some in-fills wherever there was habitable space together with the designation of some for social and public amenities. Block corners as depicted in Figure 8.5 were co-ordinated using conventional methods from control points connected to the National Control Network. The side measurements were measured as they existed and booked as shown in Figure 8.6. The end result is shown in Figure 8.7, which is an extract from a hand drawn General Plan with details of how the block corners were fixed.

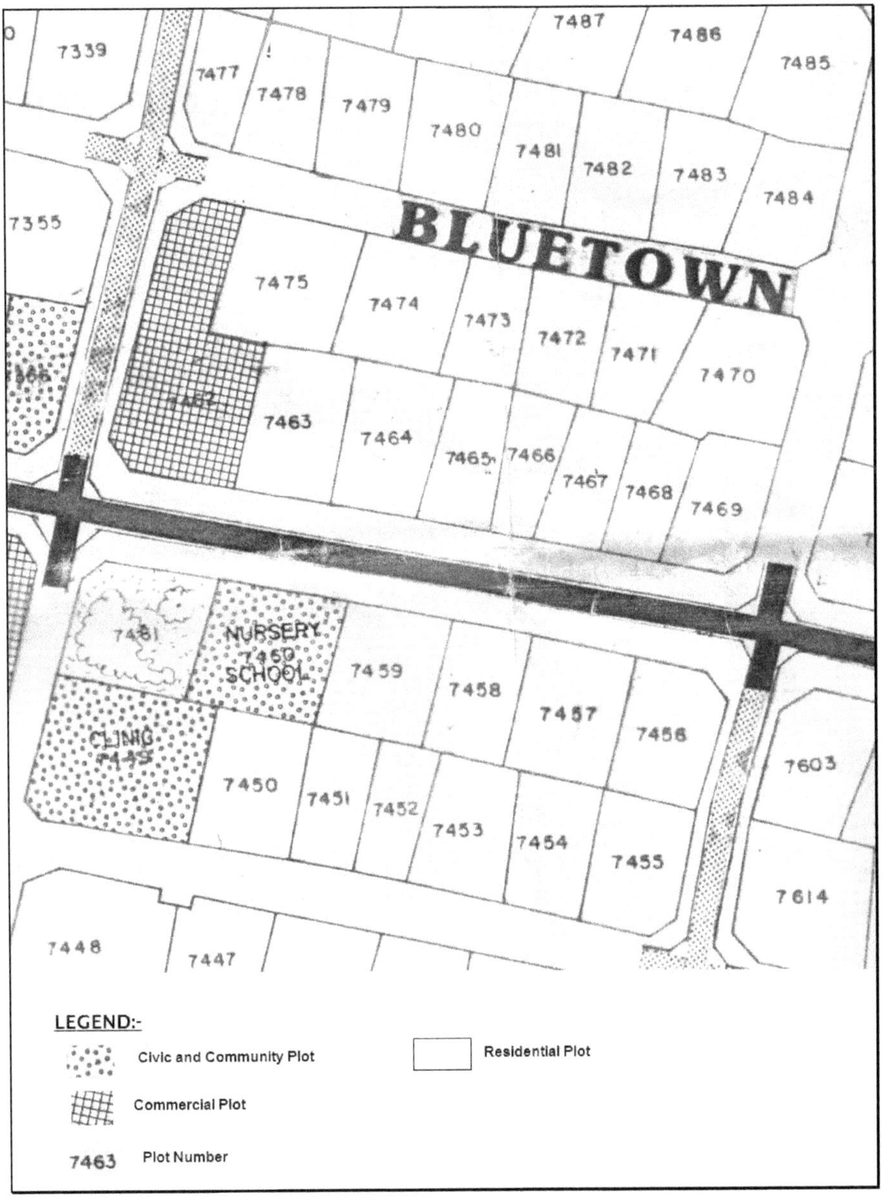


**Figure 8.4 An extract from a COR map of Bluetown**
*Source*: Department of Survey and Mapping

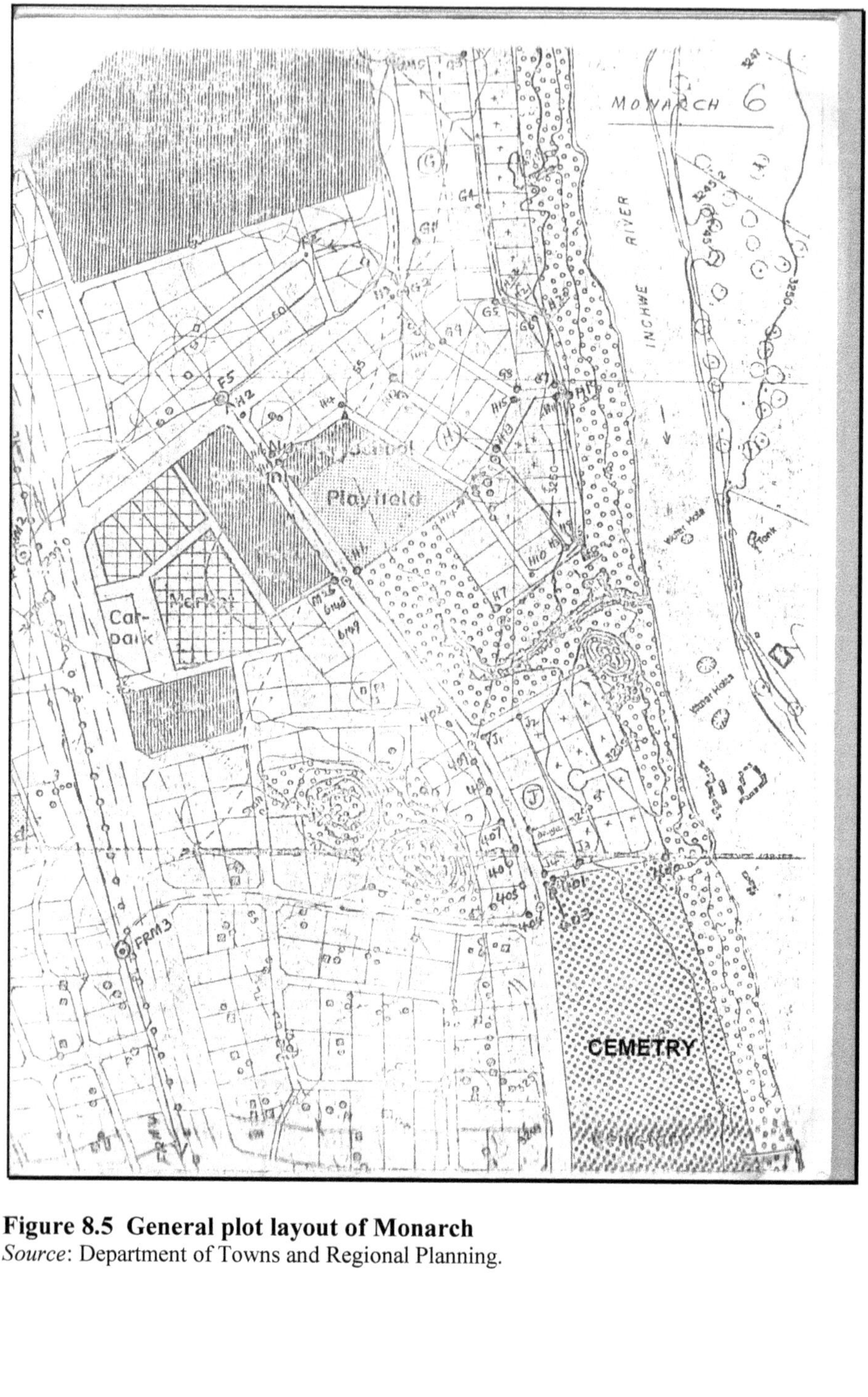


**Figure 8.5 General plot layout of Monarch**
*Source*: Department of Towns and Regional Planning.

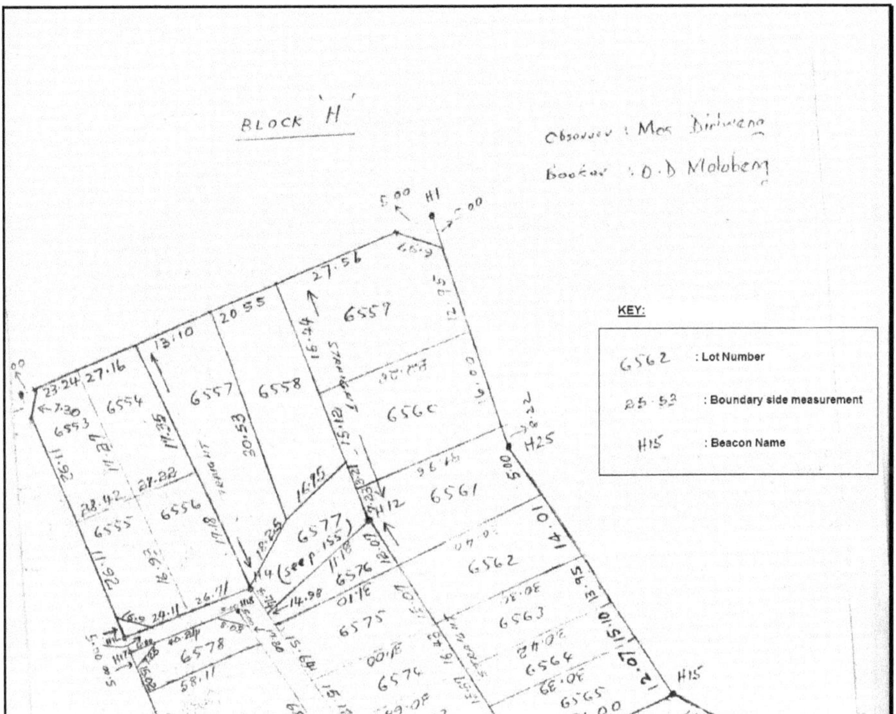


**Figure 8.6 Field measurements**
*Source*: Department of Survey and Mapping.

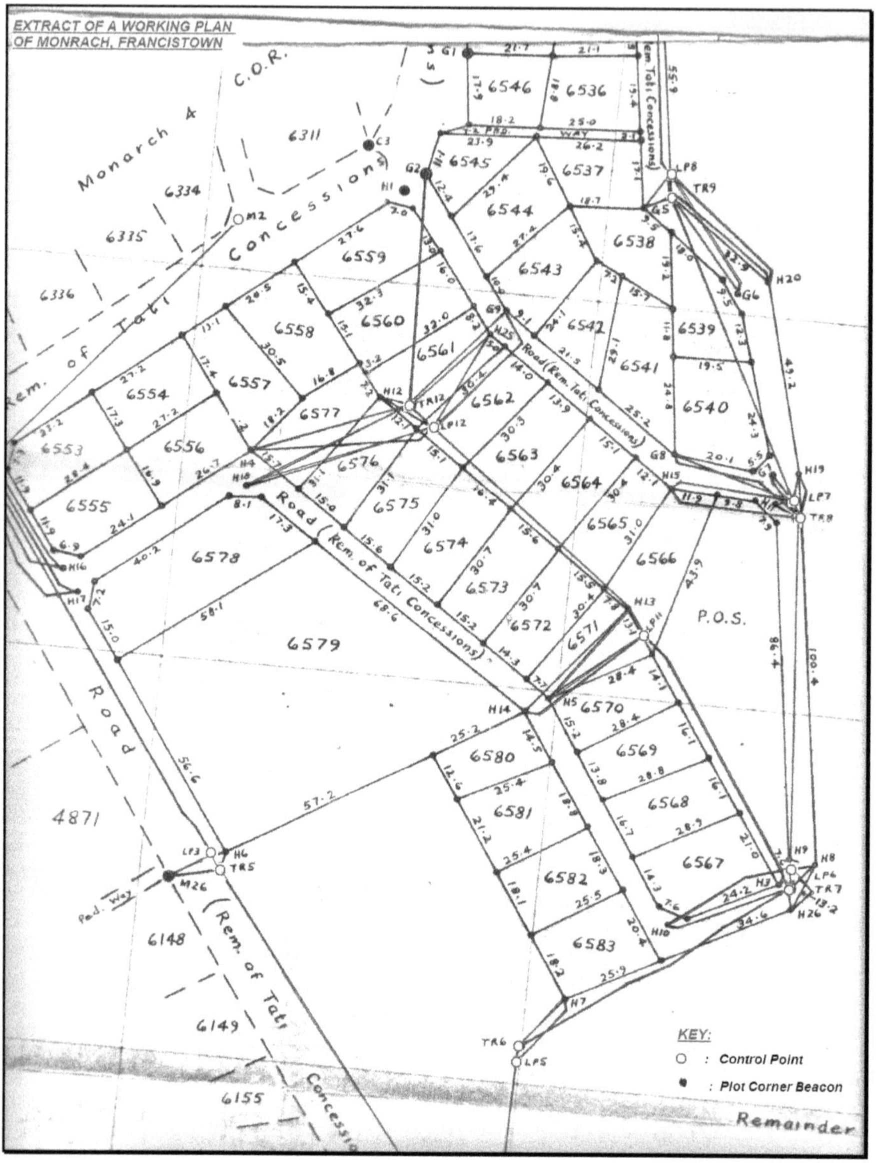


**Figure 8.7 An extract of a working plan**
*Source*: Department of Surveys and Mapping.

The informal settlements in Francistown could have been minimised had more land had been provided for reception areas, such as Monarch, which provided basic minimum standards in the provision of services. This would have legalised settlements by providing some form of tenure security. The availability of plots would have absorbed the newly arrived migrant as well as discouraging any informal settlements from emerging, which could have saved the Government and Council from the laborious and often expensive method of mapping, thinning out, relocating and then upgrading such settlements.

Through the Accelerated Land Servicing Programme (ALSP), the Government addressed the housing and serviced land supply. This was meant to meet the immediate need and demand equitably and efficiently through physical development as well as reviewing some policies and institutions for efficient implementation. The Government and Council did their best by the provision of all the necessary needs for housing but the problem always crops in the form of people who acquire land as a mean of business

## The provision of survey data for upgrading informal settlements

One of the fundamental problems involved in the registration of new plots are the regulations that defines the accuracy of the surveys that could be used. In the case of Botswana, surveyors using 'modern' survey techniques and instrumentation could result in the contravention of regulations, such as Regulation 11, which defines '*the limit of allowable error in field work*'. The laid accuracy limits cannot be attained through, for example, the use of total stations, GPS, aerial photography, and satellite remote sensing. Currently the Act is being amended to include GPS surveys and other modern survey techniques, which are widely used throughout Botswana, but not provided for in the Act.

This is particularly pertinent given that one of the major challenges is the provision of current data that will form the basis for the management process. In many instances, the only reliable mapping of informal settlements is based upon out-of-date maps, supplemented by aerial photography, and largely irrelevant census data.

Effective decisions can only be made when there is accurate information and the initial stage in the process of managing the informal settlements must be the compilation and integration of information from a wide range of sources in a transparent manner. Typically, combinations of satellite imagery and aerial photography could be georeferenced and mosiaced to provide a seamless image of the area, with the informal settlements highlighted, using a GIS framework. From this, it would be possible to detect changes between the older maps and the more recent imagery and thereby identify where the new settlements have developed and which areas have increased in size. The potential role of remote sensing in this, and the more general context of land management, offers many opportunities especially with the development of commercial, high-resolution sensors although, conversely, the data provided by those sensors have led to further problems in the interpretation and management of those data.

**Findings and recommendations**

There are certain factors and roles that can heavily influence the success or failure of upgrading projects. The most important is the acceptance and willing participation of all involved or have an interest, the relevant central and local government departments, service providers, people living in the informal settlements, construction companies, non-governmental organisations, and the funding institution (if the funds are not being availed by the government).

It should also be noted that good schemes look great on paper, as they tend to dwell on how the informal settlements are going to benefit, these being alternative employment, training as to gain new trade skills, provision of infrastructure, legal framework, development policies, and many others. As is evident from many examples, few are actually implemented which results in the creation not of a new location envisioned as a city for all, but a city for a few.

Upgrading of informal settlements is dependent on many factors, which differ from country to country although the cause is always similar: rural to urban migration which maybe also influenced by many factors, for example drought; for economic reasons; refugees due to civil wars and many others. The procedure, which is followed for the formalisation of the settlements, is dependent on the affordability of the target community with regards to service costs and the costs incurred in implementing the project.

Many developing countries are still using evictions as a deterrent to the mushrooming of informal settlements around urban areas. This includes Botswana where bulldozers were recently used to destroy a settlement within the Gaborone planning area (Nkepe, 2004), which was against government policy of providing all citizens with adequate housing without destroying the social and economic bondage of any community (GOB, 1998).

For the Francistown upgrading exercise, the communities benefited as they secured tenure over their plots and the government provided funds through which they could build durable houses. The design and implementation of the infrastructure has been stalled in Monarch by the irregular shape of the plots and as such, some plots will have to removed and those affected, given plots elsewhere.

There is also environmental degradation due to effects that could have been considered with prior planning. This includes the pollution of underground water from public and private wells. The main sources of pollution were identified as pit latrines, mine waste dumps, and landfills. New infrastructures, such as the recent alignment of the sewer pipelines along riverbanks have not only destroyed the natural flood barrier but have potential in serious water pollution.

The historic gold mining has left hazardous legacies such as underground mine workings, which have never been mapped. These are liable to gradual subsidence or even collapse, which may cause fatalities as most of the built-up areas are situated on top of them, an example being structural damage caused to the Nyangabgwe Referral Hospital in 2000. In 2003, deep holes developed in Monarch. From these examples, there is a clear need to thoroughly investigate for any hazardous conditions prior to any development by using geophysical methods. The mine dumps, which are evident in many parts of the city, contain high quantities of toxic pollutants, which pose problems to both people and foraging animals.

## References


CSO, 2003. *2001 Population and Housing Census*, Department of Printing and Publishing, Gaborone, Botswana.

De Sampaio, M.R.A., 1994. A case study of 'Favelas' in Sao Paulo, Brazil, *Habitat International*, **18**(4), 81-97.

Dickson, W.L., 1982. Land tenure and the developing world, *Land Policy and Agriculture in Eastern and Southern Africa*, Gaborone, Botswana. http://www.unu.edu/unupress/unupbooks/80604e/80604E05.html(accessed 11/03/2004).

GOB, 1982. *National Housing Policy*, Government of Botswana, Government Printer, Gaborone, Botswana.

GOB, 1998. *Botswana National Settlement Policy*, Government of Botswana, Government Printer, Gaborone, Botswana.

GOB, 2003. Government of Botswana, Gaborone, Botswana http://www.gov.bw (accessed 05/01/2004.

Malusardi, F. and Occchipinti, G., 2003. Informal settlements upgrading: the Gecekondus in Ankara, *39th ISoCaRP Congress 2003*, Cairo. http://www.isocarp.org/Data/case_studies/374.pdf (accessed 23/11/2003)

Majale, M.M., 1998. Basic infrastructure in informal settlements in Kenya, *24th WEDC Conference*, Islamabad, Pakistan. www.lboro.ac.uk/developments/cv/wedc/papers/24/G/majale.pdf (accessed 13/11/2003).

Molobeng, O.D., 2004. *Upgrading of informal settlements – Francistown, Botswana*, unpublished BSc (Hons) Dissertation, School of Computing and Technology (Field of Surveying), University of East London.

Newman, D. and Venter, F., 1999. Gauteng Provincial Government: image processing and GIS in local planning, *ER Mapper Magazine*, **1**, 12-13.

Nkepe, S., 2004. The human side of demolitions, *Mmegi Newspaper*, **21**(49), 30th March, 2004.

Planning Commission, 2003. *Mid Term Appraisal of the Ninth Five Year Plan (1997-2003)*, Planning Commission, Government of India. http://planningcommission.nic.in/plans/mta-ch11.pdf (accessed 19th March 2005)

Pugh, C., 2000. Squatter settlements: their sustainability, architectural contributions, and socio-economic roles, *Cities*, **17**(5), 325-337.

UN-Habitat, 2003. *Guide to Monitoring Target 11: Improving the Lives of 100 million slum dwellers*, http://www.unhabitat.org/programmes/guo/documents/mdgtarget11.Pdf (accessed 19/12/2003).

Wallis, M. and Mumtaz, B., 2004. Background report on sustainable human settlements development and management, *Report on Human Settlements: CSD* 12, Development Planning Unit, University College London, Presented at the *12th Session of the Commission on Sustainable Development*, 19-30th April 2004, New York.

Chapter 9

# The Contribution of Foreign Assistance to the Development of Land Markets and the Strengthening of Property Rights: The Case of USAID

Peter C. Bloch, Susana Lastarria-Cornhiel and J. David Stanfield

## Historical context

USAID policy concerning access to land[1] and other assets has undergone substantial change over time. From the end of World War II through to the early 1960s, the United States promoted administratively determined land reform. Reforms in Japan, Korea, and Taiwan were particularly effective. The importance the United States attached to land reform is evident in the 1961 Charter of Punta del Este for the Alliance for Progress, which proposed land and tax reform as *conditions* for U.S. financial aid to Latin American countries. The land reform efforts were supported in order to widen the rural population's access to land replacing the inefficient *latifundia*, thereby improving rural incomes and the productive use of land.

In the 1970s, the interest in redistributive land reforms was based in large part on the findings of USAID's 1970 *Spring Review of Land Reform*, which concluded that land redistribution should be supported because of its social and political effects on the distribution of opportunity, power, and employment. In some countries, the emphasis was more on economic factors, while in others political factors weighed more heavily. Land redistribution to the peasantry from the latifundia sector was supported not only as a way to improve the levels of living and productivity of the peasantry, but also as a tool to forestall the rise of

[1] We use the term *land* in its broad sense, including the concepts of 'real estate' and 'immovable property'. *Land tenure* concerns the formal and informal rules by which people have access to land and influence how they use the land. *Tenure security* refers continuing access to land or related resources and is a characteristic of both customary and formal tenure systems. Under many established systems of private property, tenure security rests on legal documentation and the state's guarantee of property rights.

allegiance to communist movements. Thus, the United States supported land reforms in El Salvador and Viet Nam – but not in Salvador Allende's Chile or Sandinista Nicaragua. U.S. commitment was substantial: of 52 countries surveyed by USAID in 1980, more than half had activities attempting to correct inequitable access to land. Another third had such activities in the planning stage in that year.

In the 1980s, U.S. foreign assistance shifted to emphasise macroeconomic policy reform and private enterprise development. This shift was reflected in USAID's Policy Determination on 'Land Tenure' (PD-13) that mentioned neither land reform nor redistribution (USAID, 1986). Instead, the policy emphasised land markets, land titling, and real property registration. Policy options that relied on market forces were seen as more feasible politically than administratively determined land redistribution. In addition, evaluations showed that successful land redistribution required costly additional investments in ancillary services and rural infrastructure to support reform beneficiaries.

In the 1990s, land tenure became linked to two new concerns: environmental degradation, and the disintegration of the Soviet Union. Concern over environmental degradation highlighted, among other things, the potential conflict between, on the one hand, small farmers and landless people who needed land to produce food, and on the other hand, those who championed the need for protected areas and conservation of biodiversity. The disintegration of the Soviet Union and the end of the Cold War moved land tenure issues (especially decollectivisation and privatisation) to the forefront of the policy agenda in Eastern Europe and the new independent states. The restructuring of land management, from Albania to China, is one of the more massive economic, political, and social changes in recent history. The change combines land converted to the socialist property models after World War I with that converted to the socialist property models after World War II. Moreover, the institutions of land administration, which guide and support the private property model of land administration simply did not exist.

There often appear to be sound economic, social, and political arguments favouring more widely distributed access to land and the redistribution of property rights. However, redistribution of rights to land implies some confrontation of the landless with those whose landholdings would be affected (be they landed elites or privileged state farm managers). It may also imply confrontation among ethnic groups who compete for access to land as well as confrontation between movements favouring land reform with movements attempting to regulate the use of land.

Issues associated with reforms in land tenure are almost always politically sensitive. Investments to improve access to assets have fostered more impassioned political debate than most other donor programmes intended to accelerate economic growth and agricultural development. The evaluation literature emphasises repeatedly that the most important precondition for land reform and related investments to improve tenure security is strong political will to overcome political opposition. The economic arguments for and against land reform of the latifundia and state farm systems are usually swept aside by the political urgency of land reform, or negated by the intransigence of vested or competing interests

(Lambert, 1997). In other words, these programmes have internal political purposes and require an internal political commitment:

> One lesson of the Alliance for Progress was that financial assistance and political pressure from the outside are not sufficient to convince an unsupportive government to implement a meaningful reform (Dorner, 1992).

The issues in the 1990s concerning the mechanisms by which people have access to land have reflected the waning political pressures for land reform in existing capitalist economies. At the same time, however, there is considerable interest in mature capitalist countries in land reform of the socialist economies. Both of these trends are occurring in a context of 'global markets' having a growing influence on the rural sector. Globally inter-linked product, capital, and labour markets have increased pressures for markets in land to function dynamically and transparently. Investors want to invest in profitable enterprises around the world, and this requires secure property rights. Those with capital wish to earn income from lending activities, which requires the ability to mortgage land. People who move to places with greater opportunities require access to land. Disadvantaged people often become more marginalised when there are few, if any, safety nets or programmes designed to assist them participate in land markets. Properly functioning land markets are vital for balancing these often-competing claims for land.

The 1990s also saw increasingly severe ethnic conflicts (often land related) emerge as the discipline of the Cold War evaporated with the collapse of the Soviet Union and its control over Eastern Europe. Groups have attempted to assert their rights over land to correct what they perceive as historic wrongs, using the banners of nationalism and thereby putting land on the political stage. Associated with the historic geopolitical transformation occurring in the 1990s have been an increasing number of conflicts – often civil wars – notably in sub-Saharan Africa. Some observers (e.g., Kitay, n.d.) believe that disputes over access to land and other resources have played an important role in virtually all countries in Africa where civil war has broken out, including Somalia, Rwanda, Burundi, Angola, and Liberia. Labelling these conflicts as tribal often masks underlying conflicts over natural resources. Confronting the issue of resource rights is not only critical in responding to post-conflict situations, but also may be key to preventing such conflicts in the first place.

Responding to these challenges, donors, in particular the World Bank and IFAD, have recently brought access to assets (especially land) to the forefront of their development assistance agendas. The World Bank sponsored a 4-week Land Policy Conference via e-mail during March/April 2001, which attracted 527 subscribers. During the 28-day period, 115 separate contributors exchanged 267 messages. The Bank's draft paper on 'Land Policy and Administration: Lessons Learned and New Challenges for the Bank's Development Agenda' provided a common point of departure for the conference. Following the e-conference, the Bank and USAID jointly hosted a two-day Consultation Meeting in Washington to tackle some of the thornier land access issues and to develop consensus on next steps.

IFAD released its 'Rural Poverty Report 2001: The Challenge of Ending Rural Poverty' in February 2001. Increasing access to assets by the rural poor is at the heart of IFAD's recommendations to end rural poverty. The new administration at USAID, too, has signalled its intent to emphasise agricultural development to reduce poverty, malnutrition and hunger and to stimulate overall economic growth.

## Theoretical and methodological framework

Underlying the hypothesised relationship between land markets (and property rights) and economic growth are three propositions. The first two concern the demand for productive investments and the third deals with the supply of financial resources required for investments to be realised. Private ownership (or security of rights):[2]

- gives *owners* (or rights holders) the ability to recoup investments they make in their properties, thereby providing *incentives* for investments;
- gives *entrepreneurs* (or rights holders) the opportunity to acquire land from owners who wish to alienate by sale or lease, thereby providing *opportunities* for productive investments (Stanfield 2001, p. 5);
- enhances the ability of entrepreneurs (or rights holders) to obtain mortgaged credit to finance investments.

There is a large literature concerning these issues, which is comprehensively summarised by Carter et al. (1995). IFAD's recently published analysis of the causes of rural poverty is unequivocal when it concludes that:

> The key to sustainable rural development is legally-secure entitlement to assets – land, water, credit, information and technology – on the part of the poor. Without secure property rights, farmers lack the incentive to invest in land management (IFAD 2001, p. 71).

In short, land market programmes provide both opportunities and incentives for investment. Such programmes typically include interventions such as:

- the privatisation of state enterprises (typically in ownership but sometimes in usufruct or leasehold tenure forms);
- issuance of legal titles to land as the documentation of the privatisation decisions;
- registration of those titles in an office where titles which are recorded and displayed have legal protection; cadastres, the identification of boundaries of land parcels; and

---

[2] As distinct from *public* ownership or rights. Private ownership is not necessarily *individual* ownership.

- adjudication of conflicts over rights to land and/or over parcel boundaries. Such projects are intended to improve tenure security and enhance the marketability of land.

By contributing to the more efficient use and allocation of land, they support the goal of economic growth and agricultural development.

For land markets to function most effectively, the land in private ownership must have a title attached to it, that title must be registered in a legally sanctioned registration system, and a legal framework must support the registered rights (Stanfield, 2001, p. 8). These systems may be formal or informal. In the African context, for example, local authorities may recognise exclusive use rights and a functional system to resolve disputes that may be in place, consistent with customary law. Formal titling, registration, and legal structures can be costly and are not always required to assure sufficient security for increased land market activity.

A properly functioning land market can have at least two additional benefits from the perspective of farmers. Firstly, secure property rights can provide farmers with an incentive to protect soil and other natural resources, thereby supporting the goal of environmental sustainability. In fact, McClelland (1996) recommended that donors support soil and water conservation programmes *only* when the intended beneficiaries have secure access to land. Secondly, countries with large numbers of landless and land-poor farmers are primed for social instability, guerrilla war, and, ultimately, revolution. Land market programmes can help redress gross inequities in land distribution by improving access to land and assuring tenure security by disadvantaged groups (the poor, women, ethnic groups). This, in turn, can help reduce the likelihood of rural unrest, civil war and complex emergencies and, thus, the need for humanitarian assistance.

The logic behind the presumption that secure property rights and efficient land markets will lead to desirable socio-economic outcomes is illustrated in (Figure 9.1). Interventions shown in the left-hand box are designed to increase security of property rights, which facilitates the operation of land markets. Improved land markets, in turn, have positive effects on the economy (including increased production, incomes, equity, and sustainability). The gap between theory and practice can be large. Issues related to property rights and land markets are complex, and it is essential to understand how they interact with other aspects in a given country. Analysts must go beyond theory to identify the institutional constraints (and incentives) that can alter society's ability to ensure security of tenure and optimality of the land market operation. Moreover, analysts must possess a thorough grounding in the theory of how security is derived and how it contributes to equitable and sustainable economic growth. They must also recognise that the issues differ across countries and regions.

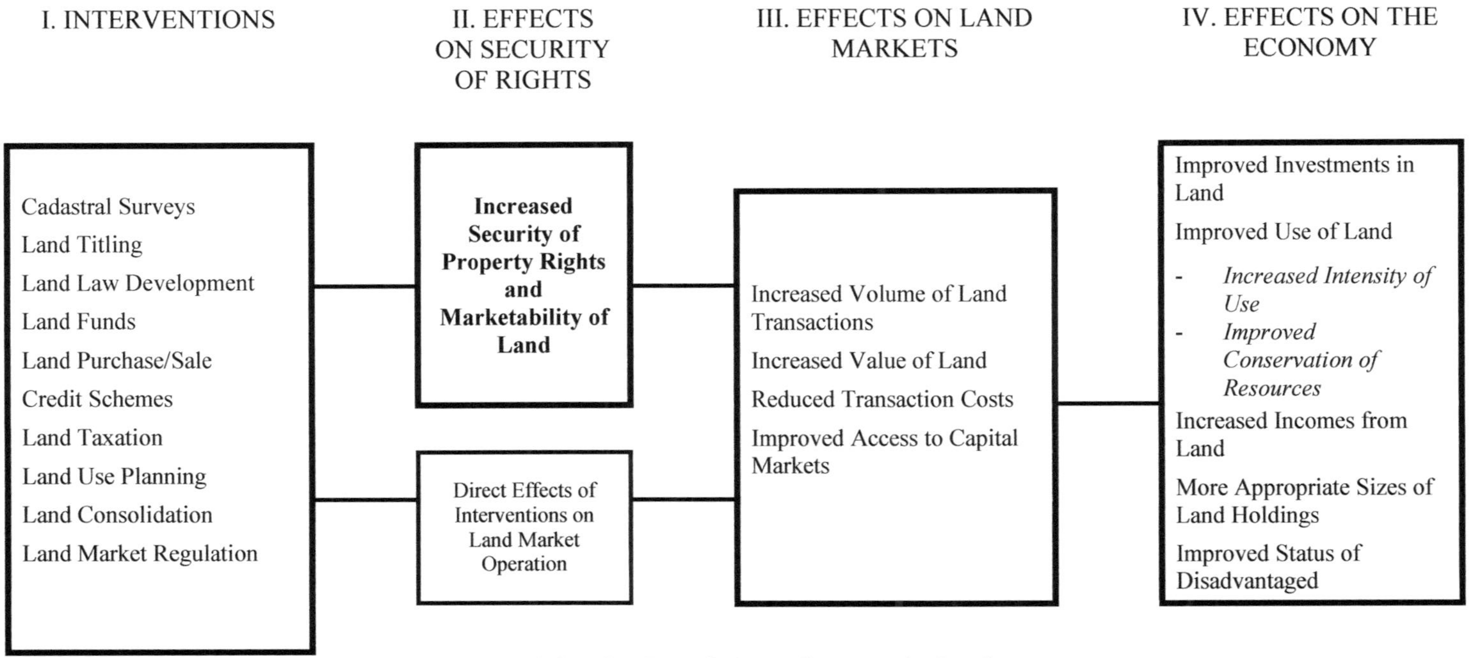


**Figure 9.1 Conceptual framework: property rights, land markets, and economic development**

Some of these exogenous factors are shown at the bottom of Figure 9.2. For example, the design of interventions may be flawed, or the Registrars may be corrupt, allowing the registry information to become contaminated with false documents and registrations, thereby compromising security. Even the best-designed land legislation cannot guarantee optimal results if enforcement of property rights and land use regulation is lacking or is strongly resisted. Even if security is attained, people may still be unwilling to access land markets to expand or consolidate their holdings. Reasons include corruption, inefficiencies in the Registration Offices, the macroeconomic situation, politics, or inadequate market information needed to make informed judgements. Even if land markets operate efficiently, these same factors plus labour and finance constraints may discourage land rights holders from risking investments in land and intensification of production. It is important to understand that these factors may mitigate the expected positive effects of land market and property rights programmes.

In addition, it is important to recognise that these programmes themselves may produce potentially negative results, such as the following:

- *Fragmentation.* Privatisation involves the division of public property into many private units, which can reduce economies of scale that may be achieved with larger enterprises. Rapid technological shifts compatible with smaller scale economic organisation are necessary as well as new forms of co-operation among these smaller scale economic organisations (Lushaj, 1995);
- *Environmental Degradation.* The shift to private or broadened ownership of land has the potential to produce negative environmental effects. Public ownership of agricultural land has been discredited, and is often identified as a feature of the previous regime in transition countries. Concentrated private ownership, as in Latin America and parts of Africa and Asia, is perceived as inequitable and inefficient. Landless or land-poor people perceive the marginal land in these large holdings to be 'open-access', and they may occupy it in ways that are unsustainable. Furthermore, transition country governments have generally neglected investments in public lands, as enforcement of exclusion has become more difficult. Moreover, the new, private owners of formerly public or large-scale private properties may decide that making long-term investments in their newly acquired properties is too risky, opting, instead, to extract immediate profit from their private assets. This may lead to over-exploitation of their assets and degradation of the natural resource base. Much of this can be avoided through proper sequencing of reforms, but it is important to recognise to these potential problems;
- *Social Polarisation.* Programmes to privatise land can create or contribute to tensions between the propertied and the property poor in several ways
  - the transfer of state owned assets to private owners in transition countries is only superficially similar to such transfers in capitalist countries. The predominance of state bureaucracies in transition countries produces

opportunities for people in positions of bureaucratic power to get disproportionate access to the privatised properties (Frydman et al., 1996);

- in some countries, privatisation has favoured or threatened to favour one ethnic group over another (Gleason, 1993; Stanfield, 1996);
- imperfections in the capital markets of transition countries may create problems in land markets (Carter and Mesbah, 1993). When lending institutions begin to function, lending may be restricted to short term, high yielding projects for a few borrowers. On the demand side, poorly functioning capital markets means that only people with accumulated capital of their own, frequently from expatriate remissions or from 'illegal' activities, can participate in land markets;
- the danger is that a few people or companies will manage in one way or another to take advantage of their position, ethnicity, or restitution claims to acquire a disproportionate share of privatised properties or to get disproportional access to capital. This will suppress operation of the land markets as they emerge, setting the stage for the polarisation of society into a few landlords and a mass of people without property.

The proposition that motivated this review of USAID's pro-poor land market and property rights programmes is, 'a key to sustainable rural development is legally secure entitlement to land on the part of the disadvantaged'.

In order to determine whether foreign assistance is contributing to improve tenure security, there are three sets of issues to explore. The first of these are the concepts of information needed for judging the success of land markets, and property rights activities. Secondly, the policy and institutional factors that affect the success of these activities and, thirdly, the implementation techniques and approaches that affect activity success. Based on these issues, the following key questions guided the review:

- what are the causal connections between investments in land markets and property rights on the one hand and, on the other, economic growth and agricultural development, especially for disadvantaged sectors, environmental sustainability, and conflict prevention?
- how do policy and institutional factors determine, a priori, the feasibility of implementing programmes to improve land markets and protect property rights? That is, what are the necessary and sufficient institutional transformations – legal, informal, cultural, and economic – that must accompany programmatic initiatives to facilitate the development of markets in land and related assets?
- assuming an appropriate policy and institutional framework, what key operational issues tend to hamper (or facilitate) implementation of programmes to improve land markets and security of tenure? That is, what specific steps and procedures are essential – both in their presence and in their sequencing – to ensure that land and related assets are securely held and become 'marketable'?

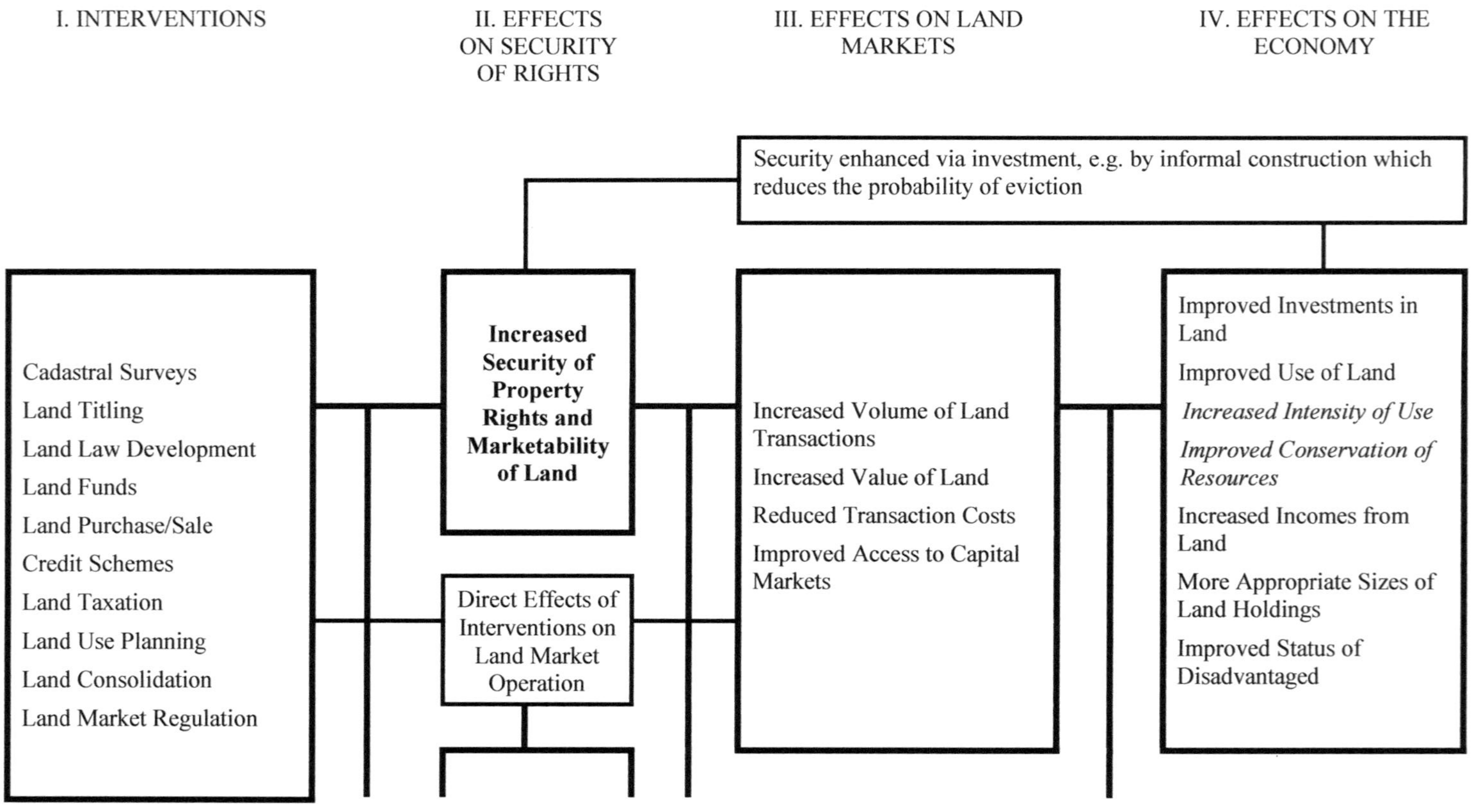
I. INTERVENTIONS
II. EFFECTS ON SECURITY OF RIGHTS
III. EFFECTS ON LAND MARKETS
IV. EFFECTS ON THE ECONOMY
Security enhanced via investment, e.g. by informal construction which reduces the probability of eviction
Cadastral Surveys
Land Titling
Land Law Development
Land Funds
Land Purchase/Sale
Credit Schemes
Land Taxation
Land Use Planning
Land Consolidation
Land Market Regulation
Increased Security of Property Rights and Marketability of Land
Direct Effects of Interventions on Land Market Operation
Increased Volume of Land Transactions
Increased Value of Land
Reduced Transaction Costs
Improved Access to Capital Markets
Improved Investments in Land
Improved Use of Land
Increased Intensity of Use
Improved Conservation of Resources
Increased Incomes from Land
More Appropriate Sizes of Land Holdings
Improved Status of Disadvantaged

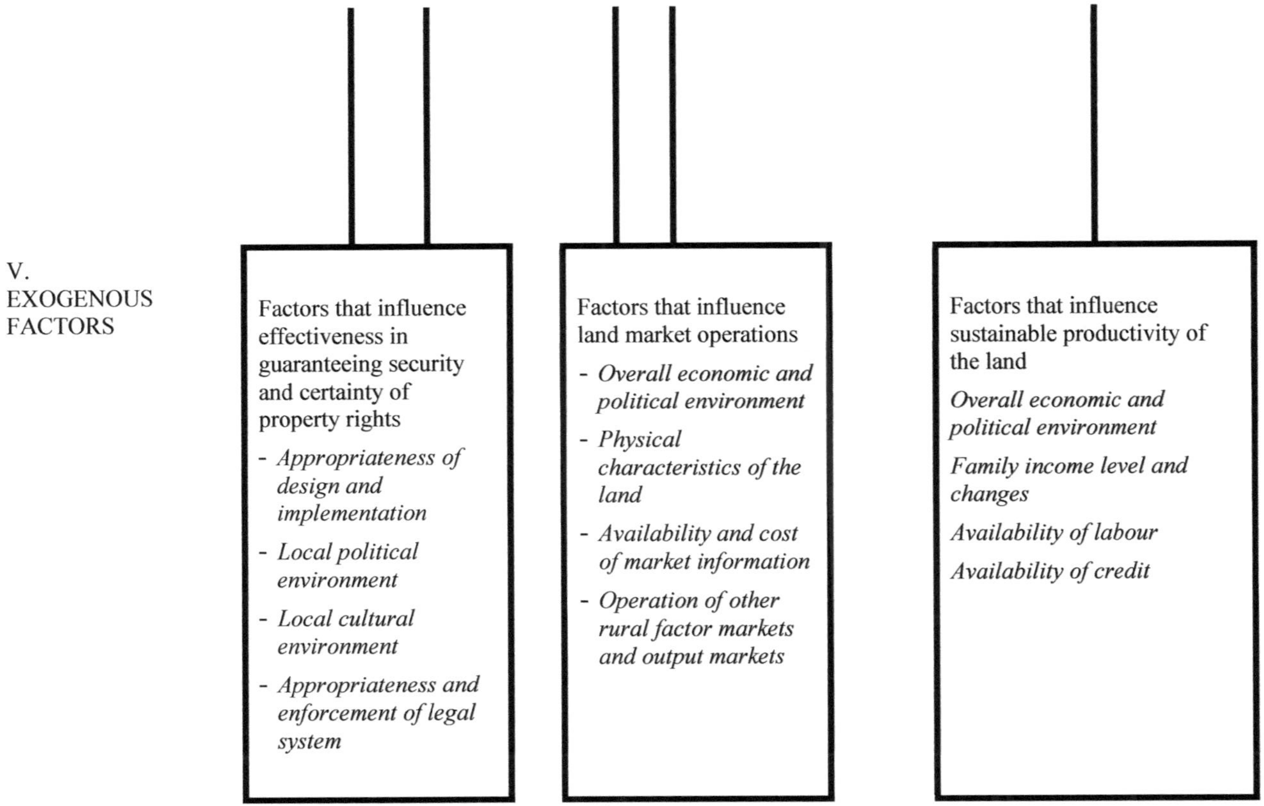


**Figure 9.2 Interventions and effects on property markets**

In order to answer the questions, a comprehensive list was drawn up of USAID projects in all four USAID regions that had significant activities relevant to property rights and land market promotion. From this list, a subset of 28 projects in 18 countries[3] – particularly those that focused on titling and registration – were analysed in 18 Country Assessments based on a review of USAID documentation found in the Development Experience Clearinghouse (DEC).[4]

The Assessments found that most USAID investment since the early 1980s have been made in privatisation programmes, titling and registration programmes (EandE and LAC regions), and land administration programmes (EandE and LAC regions). The findings based on USAID documentation were minimal with regard to attainment of project objectives and targets, and non-existent with regard to their long-term impact on land market development and socio-economic development. The paucity of findings was due both to the lack of project documentation, particularly end-of-project reports and final evaluations, and to the quality of information provided in the reports that were available.

After the completion of the Assessments, we selected eight countries[5] for more extensive review in the form of Country Briefs based on secondary sources available in the University of Wisconsin-Madison library system or on the Internet.

**Analysis based on secondary sources**

This is from information appearing in the Country Briefs and the Assessments. The Briefs are based on an exhaustive search of published literature in English, French, Spanish and Russian. The Land Tenure Center (LTC) team drafted the briefs and circulated them for comment in March 2003. The final versions are posted on the USAID website.[6]

*Analysis by region*

*Sub-Saharan Africa (SSA).* Africa differs from other regions because of the strength and vitality of customary tenure systems despite colonial attempts to individualise land in some countries, and constitutions in other countries that declare land to be owned by the State. While rights to most land are de facto

---

[3] By region, the countries chosen were: *Sub-Saharan Africa (SSA)*: Gambia, Guinea, Guinea-Bissau, Mozambique, Niger, Uganda; *Latin America and Caribbean (LAC)*: Costa Rica, Ecuador, El Salvador, Guatemala, Honduras, St. Lucia; *Europe and Eurasia (EandE)*: Albania, Armenia, Georgia, Moldova, Ukraine; *Asia and Near East (ANE)*: Philippines.

[4] Projects that supported the strengthening of community or group rights to land for improving natural resource management without reference to land market improvement were not considered (with the exception of projects in Sub-Saharan Africa) but should be the subject of a separate review. For the DEC, see www.dec.org.

[5] Albania, El Salvador, Guatemala, Honduras, Moldova, Niger, Philippines and Uganda. A ninth Brief had been produced for St. Lucia prior to the review of USAID documentation.

[6] www.usaid.gov.

private, they are not necessarily individual private property rights but rather community rights governed by customary norms.

Under customary tenure, the community has overall rights to land and the resources on the land; households and individuals hold rights derived from the community to some land. Land transfers and transactions are usually not done through a land market but rather through allocation by community authorities, inheritance, or temporary transfers such as borrowing and 'rentals' with symbolic payments. This does not mean that land transactions based on market value do not occur in the region – in highly commercial agricultural and peri-urban areas, market-based land transactions do take place. Land administration institutions were established during colonial rule and were designed to deal with those relatively small areas under freehold tenure or with state-owned land. The titling and registration of individualised private land has not been generally regarded as an urgent issue.

In general, USAID-sponsored land-market and land tenure activities in sub-Saharan Africa over the past 15 years differ from those in other regions. Typically, projects in sub-Saharan Africa:

- address natural resource management more broadly, as opposed to focusing more directly on titling/registration and land-market programmes in particular;
- often promote clear rights and responsibilities with regard to land use rather than land markets; and
- often emphasise community natural resource management within customary tenure systems rather than the granting of individual titles to land.

There are two Country Briefs for Africa: Niger and Uganda. Niger was chosen because of its innovative legal and institutional efforts to reconcile customary tenure with national legislation, supported by USAID. Niger took the first steps towards formal recognition of customary property rights by enacting a Rural Code and establishing local *commissions foncières*, or land commissions, to verify and record land ownership claims. After a promising start in the mid-1990s, however, political turmoil and consequent withdrawal of international donor assistance appear to have slowed the process of implementation. The latest relevant document is from 1996, so that it is not possible to learn about subsequent activities of application of the Rural Code. Neither does the literature on Niger contribute to an understanding of the impact of the improvement in security of property rights on sustainable development.

Uganda's effort at legislative development to formalise emerging private property rights was also supported by USAID. Studies in Uganda on the land titling efforts of the 1950s and 60s, on the other hand, suggest that there is little difference in investment on land or productivity of land among the different tenure regimes (customary, *mailo*, leasehold, and freehold).

*Asia and Near East (ANE).* The principal characteristic that differentiates USAID activities in the ANE region from those in other regions is the lack of support for land market-related programmes. USAID has had projects in ANE that strengthen the property rights of groups and communities, for example establishing community boundaries for improved management of natural resources. This is certainly an important entitlement, but these projects have not shown much concern for improving the functioning of land markets. One project was identified that dealt with land entitlement and land markets; an 18-month, $50 million contribution to the Comprehensive Agrarian Reform Programme (CARP) in the Philippines.

The programme was an ambitious attempt to redistribute most of the country's agricultural land – 10.3 million hectares and nearly 3.9 million beneficiaries. Land to be transferred included state-owned 'alienable and disposable' land suitable for agriculture, other publicly-owned land, and all private land devoted to or suitable for agriculture over a maximum size of holding. USAID decided to be minimally involved in implementation because of the controversial, politically sensitive nature of land redistribution. Assistance, therefore, was in the form of budget support rather than technical assistance.

The new rights of reform beneficiaries were to be secured by land titles and registered in the Registry of Deeds, although support for titling and registration was not directly a part of CARP, and although there were sales restrictions on the titles.[7] In the meantime, other donors, especially the World Bank and AusAid, have been assisting in the development of a more efficient and transparent set of land market institutions, recognising the huge disincentives to the formalisation of transactions inherent in the current system of land administration.

The huge literature on CARP contained surprisingly little information on its impact on the disadvantaged or sustainable development, apart from a few very recent studies demonstrating modest income gains and poverty reduction for agrarian reform beneficiaries.

*Eastern Europe and Eurasia (EandE).* In transition countries such as those in EandE, the privatisation of state-owned real property is the principal requisite for the establishment of viable land markets. This is usually in combination with the creation and proper functioning of registration/cadastral institutions to record and display the private ownership and other rights to real property. Thus projects from that region have focused in some countries on privatising state-owned property (land and buildings) including the issuing of legal and registered titles to the privatised properties, and in other countries on establishing registration/cadastral institutions to facilitate the functioning of efficient and transparent land markets.

While agricultural enterprise restructuring is common in this region, these efforts have not always produced marketable land titles nor have they invested in improving land administration institutions such as title registration offices. In

[7] Under the principal component of CARP, sale could not take place during the 10-year period when the land was being paid for, nor for 10 years thereafter.

Russia, Kazakhstan and Uzbekistan, for example, reforms have tended to encourage the evolution of new farm management forms that move the collective and state farms toward corporate enterprise models. Individual farm families are issued shares in the corporation, rather than land or property titles.

USAID documentation and Country Briefs provided some level of information leading to answers to CDIE's questions in the EandE countries.

*Question 1: Evidence of success of land entitlement projects in reaching their developmental goals*

Several steps have been taken toward achieving legally secure entitlement to land and improving the operation of land markets and encouraging farm investments resulting from legally secure titles.

*Issuance of Certificates of Title* In general, USAID-supported projects that focused on the issuance of certificates of title for agricultural land have achieved their quantitative objectives. In Albania, just over two million properties were initially registered, a 99 per cent achievement of original programme goals, although only half of all properties were registered under the programme. In Moldova, the USAID-supported effort produced the privatisation and subdivision of 826 collective farms and the registration of 844,800 agricultural land parcels, although in total there have been approximately 1.6 million private agricultural land parcels created and registered. The project in Moldova was focused more on developing cost/effective methodologies for subdivision of farms and on the issuance of title certificates and their registration than on achieving precise targets of number of parcels affected. In this sense, the project did achieve its objectives.

In both countries, the focus was on the agricultural sector and the main beneficiaries of the titling and title registration efforts were the ex-workers on the former state and collective farms. The bulk of the titling and initial registration work was oriented to the disadvantaged, in the sense of the beneficiaries being very small-scale farm families. In Albania, steps were taken also to protect the property rights of women, through the concept of 'family ownership' of the land, and the insistence on wives and other adult children being required to approve any transaction. Questions remain, however, about how effective these requirements have been in overcoming the traditional patriarchal norms and practices of many rural families, and the predominant role of the eldest male in dealing with the outside world.

Attainment of quantitative goals of number of registered titles, however, has not always been matched by the quality of the titles issued. In Moldova, at least 7 per cent of the titles are likely to have serious defects; a new $3.5 million project has been launched to detect and correct these errors. In Albania, it was estimated that about 15 per cent of the registered titles are likely to be defective in some way. Errors in initial registration discovered in 2001 in Albania are probably similar to those in Moldova, due to contractor misreporting of work done which led to the initial registration of a large number of 'fictitious' parcels. Unfortunately, little effort has yet been made to identify and correct these errors.

*Land markets* In the region, land markets are just beginning to function. Moreover, the proportion of transactions that do occur and that are being registered is less than expected. In Albania about 0.3 per cent of all properties were legally transacted in the year 2000. In that same year, there were sales of about 0.2 per cent of privatised agricultural land parcels. In Moldova registered sales, mortgages, and leases in 2001 involved approximately 1.5 per cent of all private properties. In Georgia, the cumulative total of all registered property sales transactions between 1997 and 2001 reached about 0.5 per cent of all private properties created since 1992. As a point of comparison, in 2001 in Latvia, 8.7 per cent of property units were legally transacted (Adlington, 2002). These transactions represent a small proportion of the total number of privately owned properties, but at least formal land markets are beginning to function.

An important phenomenon that deserves attention, however, is the number of informal transactions that do not pass through the registration offices. In Albania, in informal settlements, this proportion is almost over 50 per cent and near 100 per cent in some settlements, based on initial studies near Tirana (Jazoj and Kelm, 2001). While there are no data for informal transactions in the other countries, the proportion is likely to be quite high. This conclusion is based on the observation that natural life cycles for the group of new owners (frequently elderly men) should produce high rates of transmissions of title due to death of the owner. Most of these necessary changes of ownership are not being captured by the formal registration systems, let alone capturing sales and gift transactions. In peri-urban areas, the informal transaction rates where neither the buyer nor seller has a registered title appear to be as high as in the formal market transactions involving registered titles. Legal security of tenure is not a necessary and sufficient condition for dynamic land markets under certain conditions. The exact role of tenure security over the long term for influencing the dynamism of land markets needs further analysis.

Renting of agricultural land parcels seems to be of increasing importance in the region. In Moldova, for example, one study estimated that the majority of owners of agricultural land parcels have entered into lease agreements with 'peasant farms' and corporate farms,[8] which often manage several hundred hectares, most of which is leased. Such arrangements focus on the consolidation of land parcels not from a legal but purely economic point of view. The cultivation of large areas and the production of larger quantities in an organised way allow the peasant farms to achieve economies of scale intended to raise the incomes of both the farms and the lessors. Such arrangements also facilitate delivery of agricultural products to processing firms, which function very successfully in Moldova. Whether the benefits of larger scale agricultural enterprises are significant is not clear, since around half of privatised corporate farm enterprises showed losses in

---

[8] About two-thirds of the beneficiaries of the land share distribution entrusted their land shares to enterprising persons willing to manage the land and assets of a group of individuals, forming approximately 1,000 corporate farms and controlling about 46 per cent of the agricultural land.

one study. Also it is not clear whether the managers of these farms are operating in a transparent way with the lessors, showing them what profits there are.

*Investment* There are no time series data in the EandE region concerning investment on formally titled and registered parcels compared to that on unregistered parcels. One Albanian study[9] conducted at the beginning of the titling and registration process noted differences in perceived security of tenure between registered owners and non-registered owners, but very little investment by either group. From a more macro perspective, in Albania the value of agricultural production has been increasing in recent years. In Moldova and Georgia the agricultural sector, particularly the crop sub-sector has been declining since the reforms, with some signs of improvement in Moldova in the recent two years. The macroeconomic picture is distinctly mixed from country to country. Factors such as output markets, credit access, and the prices of agricultural products are important determinants of agricultural growth, in addition to the security of tenure provided by registered titles.

*Question 2: Identification and assessment of policy and institutional factors that affect success*

In most countries, the legal framework to support legal security of tenure on newly privatised and titled land is mostly in place. The USAID-supported programmes have focused on this aspect of the consolidation of the transition. Land market professionals (brokers, valuers, surveyors, lawyers, and notaries) have become established, with some successes in establishing their professional 'self-regulating organisations'. However, the real property registration offices have proven to be weak in most countries, with bribery of officials and lack of professional staff, leading to the introduction of fraudulent information into the registration systems and the gradual degradation of registration information due to inappropriate land records management.

*Question 3: Effect of implementation techniques and approaches on project success*

USAID-supported titling and registration projects in EandE countries have used innovative approaches combining land information technologies (including surveying methods) with practical approaches to dealing with lack of electricity and technically trained staff in regions outside of the capital cities. The combination of organisation and technology adaptation has led to mass titling and registration in short periods of time at low costs per property registered. However, issues concerning the quality of work done have yet to be addressed adequately.

Country Briefs were prepared for Albania and Moldova, and these two countries had similarities and differences in terms of implementation approaches.

---

[9] See Lemel 2000.

The Albania programme represents a broad focus on developing land market institutions. The privatisation of land was done by Albanian local entities, under special legislation developed for that purpose. The initial registration of 2.1 million immovable properties, urban and rural, private and public, was done by a specially chartered semi-public agency, through which Government, EU, USAID, and World Bank funds were channelled.

The Moldovan effort evolved into a comprehensive programme for the privatisation of land parcels and their subsequent registration, with minimal efforts to develop property and land market institutions.

*Latin America and the Caribbean (LAC)* Country Briefs were prepared for El Salvador, Guatemala, Honduras, and St. Lucia. These countries share a history of structural inequality and inadequate institutions, concentrated land ownership,[10] high levels of landless, land-poor families, segmented land markets, and inefficient land administration institutions. The principal focus of USAID land market interventions in those countries, with the exception of Guatemala, has been on implementing titling and registration programmes and in modernising registration/cadastral institutions. Often, the need to provide assistance in modernising and making these institutions more efficient was discovered when attempting to title and register landholdings in an efficient, timely, and transparent manner. In Guatemala, USAID programme funding focused on developing a land market by providing financial assistance to an NGO that had established a private land fund for buying large landholdings and selling family-sized parcels to landless and land-poor rural families.

Titling and registration programmes were developed to provide greater security of tenure for small farmers already in possession of land but without legal title. In Honduras, where the occupation of municipal and *ejidal* lands was common, titling initially targeted these parcels in coffee-producing areas in the 1980s, later shifting titling efforts to agrarian reform beneficiaries who had received land parcels from agricultural co-operatives. In El Salvador, USAID also supported the settlement of former combatants and the parcelling of agricultural production co-operatives and titling the land that individual beneficiaries were assigned. In St. Lucia, reforming land administration institutions, and adjudicating all land parcels, mapping them, and recording them in the new Torrens title registry was seen as a means of activating the land market. One of the perceived obstacles to land market transactions in St. Lucia has been the existence of a significant amount of land held under family land tenure.[11] The titling and registration effort was supposed to convert family land into individualised private property, family land trusts, or co-ownership with identified co-owners.

---

[10] Gini coefficients for the four countries range from 0.61 (1970s) in El Salvador, to 0.78 (1974) in Honduras, to 0.80 (1996) in St. Lucia, to 0.85 (1979) in Guatemala.

[11] Family land is land held *pro indiviso* by several heirs. In St. Lucia, many family lands had come down several generations resulting in many heirs, all of whom may not be known.

In Guatemala, USAID made a major investment in an innovative land fund in the 1980s that was to provide access to land for landless families. A private foundation purchased land, parcelled it, and then sold it under favourable conditions to beneficiary families, while providing technical assistance as these families attempted to establish their new enterprises. Since the large landholdings acquired for parcelling and resale were purchased outright by the foundation, the amount of land and number of families who could be incorporated was very limited. Another limitation encountered by this programme was the difficulty experienced by rural families in paying market prices for land and production credit. When it came to titling and registering these newly acquired parcels, the programme found a number of obstacles in dealing with the property registry; Government has to date not made the necessary changes in its registration and cadastral institutions to provide the new landowners with legal and registered titles.

A nagging problem for this review was the lack of time-series data (with the exception of Honduras) to show whether landless and land-poor families had been able to participate in that country's land markets as a result of USAID's and other agencies land market programmes.

In Honduras, four years after the titling programme began, titled farmers did not engage in sales more than the non-titled, although this period of time is very short for observing changes in market transactions. However, even in that short period, research showed that titled farmers were more likely to arrange credit than non-titled farmers. In a study conducted 10 years after the titling programme in Honduras, using the same panel of farmers interviewed in 1983, it was found that a higher proportion of titled farmers had obtained access to credit than the non-titled farmers. Credit access, however, was limited for both groups (29 per cent versus 22 per cent). Titled farmers exhibited higher fixed capital investment per capita than untitled farmers in a multiple regression analysis, controlling for age and education of the head of household, land size per capita, being a coffee producer, receiving credit and having electricity. However, there was little evidence that titling affected overall farm income per capita.

The Honduras studies provide evidence of the impact of titling on credit access and investment. However, credit access and market access continue to favour the larger farms, in a country where the distribution of land is highly skewed – most of the good land is owned by a few landholders and most of the farming families have little land. Moreover, there have not been other studies that attempt to measure the long-term impacts of these interventions on the land market, and on the participation of land-poor rural families in the land market. Land ownership in the four countries continues to be highly concentrated, landlessness remains high, and rural poverty is still a serious problem.

Monitoring of project effects has been deficient. For example, no studies were found on whether land purchasers in the Guatemala land market programme were able to finish paying for their land and whether they received registered titles for that land. Nor were any studies found in the countries where titling and registration were implemented that systematically investigated the incidence of subsequent registered and unregistered transactions, and analysed why many titleholders do not register these transactions. More importantly, there have been few studies on whether the land-poor's ability to participate in the land market has improved. It is therefore not possible to assert that these interventions have improved the functioning of the land market in these four countries, that they have improved the land-poor's access to land, or reduced rural poverty. Certainly, the Honduras studies point out positive effects of titling on credit access and investment in fixed capital in coffee producing areas, which are important findings. It seems, however, that these two positive indicators are generally limited to the wealthier segments of local communities. It also appears that access to credit through banks is dependent on the policies of individual banks and the orientations of the bank managers at the local level. Where banks do not actively embrace formal titles as evidence of ownership, the titles themselves make little difference.

A major problem in the LAC region is that land markets remain highly segmented, with only limited access to land across social strata. In addition, it has become apparent that some landholders, for economic, social, and/or cultural reasons, prefer tenure forms other than individualised private property. In St. Lucia, the amount of family land after the termination of the project remained high (25-33 per cent), reflecting landowners' reluctance to do away with a type of landholding they perceive as beneficial, even though it is neither easily marketable nor economically profitable. In Guatemala, many people in indigenous communities prefer communal land rights, and not individual title.

Another cultural/social factor is that many landholders felt secure in the occupation of their land even without title (for example, in St. Lucia and Honduras). More troubling is that many new titleholders are not registering subsequent land transactions, suggesting that for many rural families, Property Registry records are not widely valued as a source of tenure security. It would seem, therefore, that other factors, besides titling and registration, are needed for tenure security and for ensuring that smallholder families are able to participate in land markets and to benefit from them.

*Analysis by theme*

As the previous section indicates, the types of land-related projects, the state of land markets, and the impact of USAID projects on land markets and socio-economic development vary widely among countries and regions. This section presents the findings across all regions with regard to the projects' impacts on land markets. The secondary literature reviewed in the Briefs contributed a great deal to the understanding of the context and history of land tenure systems and their relationship to the broader economic and political experiences of the countries. However, it contained little evidence of the impact of the donor interventions in land markets and property rights. The section below provides more detail on the issue of documentation of achievements.

*Privatisation, titling and registration of land* Initiatives promoting privatisation were undertaken in the EandE region and in some countries in Africa. Most of these initiatives developed significant legislative reforms. For example, all programmes in EandE had substantial input into the legislative and regulatory framework being created to support and guide the operations of land markets. Similarly, all privatisation projects in that region attempted to launch strong programmes of public awareness and education concerning the new system of private property and the rights and responsibilities of the new owners of land and buildings. In Africa, USAID programmes supported legislative reform in Uganda, Mozambique, Niger, and Guinea-Bissau that, while not generally privatising land, set the legal basis for restructuring and recognition of local community and individual land rights. There was direct support to the privatisation effort in Moldova that began as an intense and rapid campaign for the privatisation of agricultural land, combined with the registration of the titles issued. In the Ukraine, the programme focused on the restructuring of agricultural enterprises and on political and legal issues.

USAID-supported activities in Eastern Europe also focused on land registration, particularly on privatised parcels of agricultural land. The Albania programme of land market development registered both public and privately owned properties, urban and rural, while in other countries the USAID effort dealt only with privatised agricultural land parcels. The USAID projects also developed procedures for the first registration of privatised land and the registration of subsequent transactions that were applied to all properties, agricultural and urban. To an important extent, this initial registration work contributed to the correction of errors committed during the privatisation process. Also included in the Eastern European and Eurasian region projects was support for the development of land market professionals (brokers, surveyors, assessors, and notaries).

Quite a few of the projects in LAC also implemented titling and registration, initially as discrete projects and later as part of larger agricultural support programmes. In general, titling and registration in LAC were not part of a privatisation process but rather attempted to formalise land ownership rights of smallholders and agrarian reform beneficiaries already occupying and working agricultural land.

The Briefs showed that projects dealt with a variety of titles, not just individual titles. In some African countries, there are community entitlements to relatively large pieces of land. In St. Lucia, inherited land was titled without subdivision to descendants of the original owners. Ownership of agricultural land was granted to families in Albania and to households in Georgia, although titles were issued in only one person's name.[12] In some countries, such as Albania, community titles to some forestlands have been granted. Such titles require special legal treatments to enable low-cost land transactions and to protect the security of landholders.

---

[12] This was true as well in the three Central American countries. This is a frequent issue in land policy. The household is frequently assumed to be a cohesive unit, with the household head making decisions (including land use and transactions) that reflect the views of all the members. There is abundant research evidence (see, for example, Lastarria-Cornhiel, 1997) that shows that household heads often act in individual self-interest.

In the EandE region, the distribution of privatised agricultural land has been quite egalitarian, especially in comparison with the older agrarian structures of Latin America and the Caribbean. The concern in the EandE countries is that the emerging farming enterprises are not succeeding economically for reasons other than their ownership status. Similarly, in the LAC region, the investments in titling of smallholders have targeted a disadvantaged sector, the small farmers, but the overall economic outlook for agriculture has not been good. Furthermore, the structural advantages of the large landholders for credit and market access appear to undermine the abilities of the small farmers to be successful regardless of their tenure status.

*Registration/cadastre institutions* Many projects in both EandE and LAC made major investments in establishing or reforming registration/cadastre institutions that display evidence of the legal owners of land and buildings. In the EandE region in particular, new Registration/Cadastre institutions have been established while entities that had functioned for recording the occupation and use of land and buildings under the previous regime had to be closed or adapted to new conditions. Programmes in Albania, Armenia, Moldova, and Georgia made investments in newly established systems for the registration of titles and land transactions, and for the mapping of property boundaries, including major investments in the computerisation of these land information management systems. The approach in LAC has been to reform, modernise, and link registration and cadastral (mapping) agencies in order to make them more efficient and accessible to all landholders and to facilitate land market transactions.

In EandE and LAC, efforts were made to make these registration/cadastre agencies accessible to the population and as efficient as possible. Registry officials, however, often have taken advantage of their control over property records to extract bribes from clients for expedited service or for manipulation of those records thereby producing serious threats to the integrity of the Registries and to the efficiency of their functioning.

In both regions, there appear to be a significant proportion of transactions that are not being channelled through the Registration Offices, which indicates either higher than acceptable transactions costs (including taxes, 'facilitation fees', time) or a lack of public appreciation of the value of registered title. Serious consideration should be given to re-thinking the institutional design of the registration/cadastral system in order for the investments in titling and initial registration to be durable.

*Impact on land markets* The evidence on whether entitlement programmes produce more dynamic land markets is mixed. One objective of entitlement programmes is the clarification of title to facilitate access to credit. Studies in the LAC region that compared titled and untitled landholdings at one point in time typically showed that the holders of legal title have more mortgage agreements with credit institutions and get more loaned funds than do holders of land without legal title. These studies are instructive, but they do not demonstrate that improving legal status of titles to land 'causes' improved access to credit. Other factors may be producing both legalisation of title and access to credit, such as local bank lending policies and the amount of investments that landholders have made in their parcels, particularly in coffee-growing areas. Those who have expanded their coffee plantations may seek legal protections of these investments, and at the same time with these investments may find it easier to get loans for more investments. Longitudinal studies are needed, before and after titling.

Data gathered in Honduras from a sample of titled and non-titled landholders, before and after the issuance of new titles, showed that legal titles appeared to have an impact on access to credit. This was despite the relatively short time period of the study (four years between the baseline study and the follow-up study of the same sample). Private banks that require collateral for loans accounted for all of the additional loans, and the amount of money loaned to titled landholders, as compared to those in the (untitled) control area, rose dramatically. When it was noticed that the great majority of these loans occurred in only two communities, case studies were undertaken and revealed the willingness of a single private bank manager[13] to accept the new titles as collateral for loans in the two communities. In the communities not serviced by this bank manager, there were no significant changes either in the number of clients or in the amount of money loaned.

The experience of the innovative bank manager, however, is very instructive and could be explored further for demonstrating the value of using the new titles as he did. Local conversations about the new titles seem to have encouraged the bank manager to experiment with using the new titles for issuing loans. If the policies of other banks could be the focus of special efforts to make them more responsive to the needs of the newly titled peasant producers, then the increased importance that many farmers attach to their land titles could be leveraged into greater access to credit. It is possible that over the long term, it could begin to affect the more difficult processes of on-farm investments, increased productivity, and farmer income enhancement. Titling alone, without local educational efforts and institutional changes will likely not bring about these effects. Longitudinal surveys of landholders will be insufficient to detect such relationships; they should be complemented by in-depth, historical case studies of the communities and farmers being touched by the titling and registration programmes as well as movements toward or away from broader democracy of opportunity.

What about transactions in general? The study in Honduras compared the transactions in titled areas with transactions in untitled areas, four years after the titling programme (Stanfield et al., 1990). In one Department, there was no change

[13] The bank manager, who knew many landholders from the days when he had been the head of the local coffee cooperative, made loans based on this personal knowledge but did not register mortgages on the titled properties with the Property Registry. He simply put the new titles in his vault until the loans were repaid. A similar situation occurred in the other community where credit access dramatically increased following the titling programme (see Stanfield, Nesman, Seligson and Coles 1990).

in the proportion of parcels affected by transactions after titling in comparison with the proportion before titling. In another, the proportion of parcels that had been transferred during a four-year period was actually *less* among the titled landholdings than in the untitled area. The titling programme – the issuance of legal land titles to private owners in replacement of their customary titles – did not stimulate greater sales of land, at least in the short term. In the minds of titled landholders, however, the value of their parcels increased with the issuance of title much more than the increase in value for non-titled parcels. Actual market prices need to be examined to see whether this perception of changed value would translate into increased market values. What will occur over the longer term remains to be seen.

In the EandE region, agricultural land markets are inactive for the most part, except for agricultural land on the periphery of expanding urban areas and along highway corridors, which basically involves transactions for non-agricultural purposes. In the more stable agricultural areas, the low transaction rates may indicate that the new owners are not interested in selling what they have just acquired from the land reform. It may also be the case that the low profitability of agriculture does not tempt people with capital to wish to acquire agricultural land. There may be people who wish to sell, but there may also be few people who wish to buy. The low rates of transactions noted above in the EandE region may reflect a combination of these situations (see Adlington, 2002). Conversely, there is evidence that de facto privatisation without formal title (for example, in squatter settlements around Albania's capital city) where the demand for house plots is substantial and the cost of formally privatised land is high, a large number of transactions occur (Jazoj and Kelm, 2001).

Registered titles do not automatically produce transactions. Unregistered titles do not seem to hinder transactions. Then, why do more settled, market-oriented economies in Europe and North America require registered titles for market transactions? Probably the key term is 'settled'. In the LAC region, highly polarised societies afflicted by war and internal conflict are not 'settled'. Nor are many of the transition countries of the EandE region close to being 'settled'. It seems reasonable to try to move toward a more settled state of affairs, step by step. Settling property rights of landholders and consolidating institutions, which support them, are important steps.

## New directions for USAID investments in land markets and property rights

Based on the documentary evidence and LTC's knowledge of the issues that have arisen in the course of implementation of USAID investments in land markets and property rights, there is a need to rethink USAID strategy for the next generation of land-related development assistance. Policies and programmes to consolidate the investments made in privatisation and registration of property rights should focus on seven themes.

*Increase substantially the proportion of transactions recorded within the formal registration and cadastral institutions*

Constructing land market institutions and registering property rights in them is only the first step in the process of developing efficient, equitable and sustainable land markets. It is also necessary to ensure that subsequent transfers of rights are recorded there as well; if transactions are not recorded accurately, the registration system will quickly become unreflective of the true ownership situation. The disincentives to the recording of transactions, such as transfer tax, bureaucratic hurdles, and corruption are high in most countries. As DeSoto (2000) argues:

> [I]n every country we investigated, we found that it is nearly as difficult to stay legal as it is to become legal. Inevitably, [property owners] do not so much break the law as the law breaks them – and they opt out of the system.[14]

Many landowners appear to be conducting sales, leases, inheritances, and gifts involving land and other real estate through private arrangements, often with documents such as deeds of sale, avoiding the formal registration/cadastre agencies. In Georgia, for example, only 1.2 per cent of the 2.4 million privatised agricultural land parcels have been sold and legally registered over the past six years, a rate of 0.2 per cent per year. Considering life expectancies and the fact that most beneficiaries of those programmes tend to be the oldest family members, it would be logical to expect that at least inheritances would be occurring at a much higher rate. Sales, gifts, and mortgages together should add more. This very low rate leads to the suspicion that transactions are occurring outside of the formal registration system. Gradual degradation of the public records created by the titling and registration projects is therefore occurring, and seriously threatening the substantial investments made.

Possible interventions include:

- reduce transactions costs, including informal payments and transfer fees;
- bring land registries closer to the population, geographically, socially, and culturally;
- improve the efficiency of operations of the registries, providing rapid service in a one-stop shop;
- substitute property taxation of registered owners for transfer taxes. Property taxation on registered owners will motivate them to register transactions, which transfer their ownership to another person. Removing or reducing transfer taxes will remove one obstacle to the registration of transactions;
- provide positive incentives for registering transactions, including lower property tax rates or preferential access to other government services.

---

[14] LTC research in Latin America had made this same point in the 1980s and 90s. See, for example, Hendrix 1993.

*Encourage more dynamic land markets*

Potential sellers may prefer to keep their land idle for speculative or other purposes. Potential buyers will not affect the purchase unless they have the information and resources needed to buy land. Real estate professionals or influential people may attempt to manipulate market information for speculative purposes if transparency is lacking. Possible interventions include:

- improve the quality and broad availability of information about land owned by people who are interested in selling;
- set a property tax rate noticeable enough to provide an incentive for people who are merely holding land for speculative purposes to either put the land to economic use or to place it on the market;
- encourage financial institutions to broaden access to mortgage credit facilities, in order to provide long term capital for the purchase of and investment in land.

*Improve access to credit for titleholders*

As numerous studies have shown, titling alone, without local educational efforts and institutional changes, will likely not bring about improved access to institutional credit. The experience of an innovative bank manager in Honduras who offered credit to new titleholders may be instructive and should be explored further for demonstrating the value of titles. The local conversation about the new titles seems to have encouraged the bank manager to experiment with using the new titles for issuing loans. If the policies of other banks could be the focus of special efforts for making them more responsive to the needs of the newly titled peasant producers, the increased importance that many farmers attach to their land titles could be leveraged into greater access to credit. Over the long term, this might begin to affect the more difficult processes of on-farm investments, increased productivity, and farmer income enhancement.

*Improve the guidance of land markets*

Environmental concerns and the high costs of retrofitting public utilities and other infrastructure once settlements have formed indicate that there is a public interest in developing rules to guide land use and the location of urban development, including land use zoning and development control. Such rules require functional institutions, distinct from registration and cadastral agencies but sharing information, for their application.

Governance capacity is weak in most countries. In EandE, the centralised model inherited from the previous regimes is very resistant to the decentralisation of the governance of land. In LAC, Asia and Africa, where there has been more experience with local government models, political instability and/or highly polarised landholding patterns often convert local government into local protectors of powerful elites. Under such situations, it is difficult to conceive of governance models based on consultations with local stakeholders as well as local government officials. Centralised budget and fiscal structures frequently deprive local governments of the resources, which they need to function. Nonetheless, without much strengthened local land governance, the sustainable use of land and more equitable access to land will be difficult to achieve by relying only on the unguided operations of land markets.

*Improve the secondary uses of registration/cadastre information*

The theoretical basis of much of the land titling effort has been the private benefits expected to flow from greater security of land tenure via the adjudication of titles. These private benefits, according to the advocates of land titling programmes, should accumulate sufficiently to be of benefit to the wider society.

The data indicate, however, that the expected private benefits of titling on the welfare of the land-poor have been small and difficult to realise, at least in the short run. However, the experience with various titling and registration programmes has suggested that there may be significant social benefits from making the registration and cadastral information more widely accessible. Such information can improve the capabilities of the government, public utilities, private developers, and conservers of the land and attendant resources to administer these resources more effectively and closer to the people.

The users of this property records information come to have an interest in the quality and accuracy of the information. Pressures can emerge from these users on the registration offices to operate more transparently, at lower cost to the potential clients, and with more of a client satisfaction culture, in order to reduce the incidence of informal, unregistered transactions and to assure the accuracy of the information in the Registries. The following paragraphs outline some of the most attractive uses.

*Inventory and boundary identification of publicly owned land* Comprehensive registration information can provide detailed data concerning the distribution, use, and location of land, as well as identify the public authorities with responsibilities over public lands. This information can be extremely useful for planning development projects which affect the land, such as soil and water conservation, watershed management, technological transfer oriented toward particular crops, and programmes seeking solutions to the problem of extreme poverty due to land fragmentation.

Comprehensive titling and registration also can contribute to the clarification of boundaries of land that cannot be titled nor alienated, such as forest reserves, certain critical watershed areas along public roads, and lands assigned to municipal administration, which are subject to different rules for occupation and titling. By identifying publicly owned lands that are not subject to alienation, other lands that are potentially available for private ownership and use also are identified.

*Demarcation of boundaries of administrative units* Boundaries between administrative units are frequently ambiguous and often undefined. Many titling and registration efforts confront this problem when carrying out the cadastral mapping of parcels, which requires the identification of parcels according to their location by administrative unit. In many instances the existing maps of boundaries do not correspond with geographical features on the ground, or there is a longstanding dispute as to which department or municipality had jurisdiction over a particular area.

This confusion has meant that in these administratively ambiguous areas, the public has difficulty knowing where to go to secure such documents as birth and death certificates and even to record land transactions. Such boundaries must be clear before the local tax authorities can collect land taxes.

*Improvements in property tax collection* One of the most important products of cadastral mapping is the potential improvement of property tax collections. The collection of taxes is often based on local knowledge of who owns what land and on the propitious appearance of a property owner who wants to carry out some transaction, which requires certification from the local registration/cadastral office. Both the valuation procedures and the rate of tax collection are often inadequate; the personnel administering the programme are often poorly trained and/or corrupt, and have little information to guide them in assessing and collecting taxes. When discounting the cost of maintaining the tax rolls, it is estimated that the cost of producing the cadastral maps and property owner lists can be recovered from increased revenues in four to five years.

*Change the approach to the design and implementation of land projects*

Some USAID regional and country offices have had limited interest in programmes to strengthen property rights and the operations of land markets, and in others there has been substantial interest in these issues. In the latter cases, it would be useful to consider changes in the ways that projects are developed to deal more adequately with land markets and property rights issues. Two effective approaches are participatory development of entitlement programmes and stakeholder analysis before and during implementation. Several of the Country Studies revealed the (potential) usefulness of these approaches. For example, experiences in St. Lucia and Niger with participatory development of their property rights programmes illustrate the importance of wide-ranging and patient consultations with people and organisations, public and private, concerning property rights and land market issues and how to address them. In St. Lucia, land commissions held hearings for nearly two years, and there were special studies done before the actual project design effort was complete. Of course, for a small island country, such consultations are much easier than in large, socially and economically diverse societies. Still, despite Niger's large size and poor communications, the Rural Code Committee managed to cast a wide net and received significant bottom-up input prior to the passage of the Rural Code and the establishment of land commissions. The public education coming out of such efforts and the greater appreciation by project designers of the problems and their solutions would go far in the direction of more responsible project design and more effective project implementation. Another type of consultation should be done once the outlines of the proposed project are defined. At that stage, stakeholder analysis identifying the groups likely to lose from and hence to oppose the new system would be useful. Such analysis should result in recommendations for the neutralisation or mitigation of the opposition, possibly via small concessions. For example, in St. Lucia, the Bar Association silently opposed the simplified conveyancing procedures set up in the new Registration Law. Once the project was underway, the Bar lobbied successfully for an amendment to the law, which required that all conveyances had to be prepared by lawyers, with attendant fees. Perhaps the negotiation of a set of standard conveyance forms and fees for their preparation during the consideration of the Registration Law could have met the lawyers' concerns and yet kept the fees much lower than they actually became.

*Incorporate land issues into USAID's strategic objectives*

The proposition that has motivated this work has arisen more or less empirically, as Missions have seen the need to deal with land issues. In USAID's Strategic Objectives, the land issue is not clearly defined. Missions have to somehow insert land projects into their country programmes within the Agency's Strategic Objectives. It would be helpful for USAID to explicitly incorporate the following proposition into USAID's Strategic Objectives:

- a key to sustainable rural development is legally secure entitlement to land on the part of the disadvantaged;
- 'sustainable rural development' means the use of natural resources in ways that preserve them for future generations. 'Legally secure entitlement to land' means the social and political means for assuring landholders that their access will not be threatened in arbitrary and capricious ways. 'Disadvantaged' means the social groups who live in poverty and do not have the opportunity for getting out of that condition, such as the land-poor, ethnic and racial groups and gender identification who suffer discrimination from the dominant sectors of society;
- why is it important to target the disadvantaged explicitly? Some would argue that simply by liberating markets, including land markets, economic growth will occur in the most sustainable form over the long term and with maximum benefits for all sectors of the population, including the disadvantaged. Whether this perspective is correct is partly an empirical question, but there is enough evidence to doubt the automatic linkages between unfettered markets and improvements in the lives of the disadvantaged.

## References

Adlington, G., 2002. Comparative analysis of land administration systems, with special reference to Armenia, Moldova, Latvia and Kyrgyzstan, presented to the *World Bank Conference on Land Administration*, Budapest, March, 2002.

Carter, M.R. and Mesbah, D., 1993. Can land market reform mitigate the exclusionary aspects of rapid agro-export growth? *World Development*, **21**(7), 1085-1100.

Carter, M., Feder, G. and Roth, M., 1995. Land tenure and agricultural performance: reflection on global experience, *Transition of China's Rural Land System: Papers from International Symposium on Rural Land Issues in China*, Land Tenure Center Paper 151, Madison, Wisconsin.

De Soto, H., 2000. *The Mystery of Capital*. Basic Books, New York.

Dorner, P., 1992. *Latin American Land Reforms in Theory and Practice: A Retrospective Analysis*, University of Wisconsin Press, Madison.

Frydman, R. and Rapaczynski, A., 1994. *Privatisation in Eastern Europe: Is the State Withering Away?* Central European University Press, London.

Gleason, G., 1993. Central Asia: land reform and the ethnic factor, *RRFI/RL Research Report*, **2**(3), 28-33.

Hendrix, S.E., 1993. *Property Law Innovation in Latin America with Recommendations*. LTC Paper 149, Land Tenure Center, University of Wisconsin, Madison.

International Fund for Agricultural Development, 2001. *Rural Poverty Report 2001: The Challenge of Ending Rural Poverty*, Oxford University Press, Oxford.

Jazoj, A. and Kelm, K., 2001. *Provisional Immovable Property Registration in Informal Settlements: A Study in Kamza*, Project Management Unit, Tirana.

Kitay, M., n.d. *Land Tenure Issues in Post-conflict Countries*, USAID, Washington.

Lambert, V.A. and Seligson, M.A., 1997. *Asset Distribution and Access: Land Tenure Programmes, Promoting Third-World Development and Food Security*, Praeger, Westport, Connecticut.

Lastarria-Cornhiel., S., 1997. Impact of privatisation on gender and property rights in Africa, *World Development*, **25**(8), 1317-1333.

Lemel, H. (ed.). 2000. *Rural Property and Economy in Post-Communist Albania*, Berghahn Books, New York.

Lushaj, S., 1995. *Land Fragmentation and Consolidation in Albania*, Land Tenure Center Research Paper, University of Wisconsin, Madison.

McClelland, D.G., 1996. Investments in Agriculture: A Synthesis of the Evaluation Literature, Center for Development and Evaluation, Report No. 15, USAID, Washington.

Stanfield, J.D. and Bloch, P.C., 2002. *A Conceptual and Methodological Framework for an Assessment of USAID's Investments in Land Markets and Property Rights*, Land Tenure Center, University of Wisconsin, Madison.

Stanfield, D., Nesman, D., Seligson, M. and Coles, A., 1990. *The Honduras Land Titling and Registration Experience*, Land Tenure Center, University of Wisconsin, Madison.

USAID, 1986. *Land Tenure*, Policy Determination No. 13, USAID, Washington.

USAID, 1970. *Spring Review of Land Reform*, USAID, Washington.

Notes to Figures 9.1 and 9.2

The two diagrams in this chapter are a representation of the theory and reality, respectively, of the mechanisms by which interventions can increase the security of property rights and develop land markets. Figure 9.1 shows the hypothesized path of influence of an array of interventions: direct effects on security and marketability as well as on land market functioning, and indirect effects on economic performance. Figure 9.2 illustrates the range of exogenous factors, which can have an impact on the degree to which the interventions can be effective. The broad conclusions that can be drawn from the Figures are:

1. land market and property rights programmes impact economic performance in combination with a variety of conditions and other programmes;
2. similarly, land market and property rights interventions are necessary, but they are hardly sufficient, to improve economic performance.

Chapter 10

# Comparative Evaluation of Purchase and Subdivision Processes in Denmark, England and Wales, and Belarus

Marina Vaskovitch, Robert Dixon-Gough and Erik Stubkjær

## Introduction

For any economy to work efficiently, real property should be easily transferable to make the real estate market more efficient. However, in some countries these transfers are simple and quick, whilst in others they are more problematical and costly. This is largely dependant upon the different organisational legal frameworks of the countries, and the diversity of their institutional arrangements for property transactions, etc.

The property market is expanding over the national borders and, thus, it requires that foreign experts can understand transparent and simple rules for property transfers. Therefore, the problems of transparency and comparability of land administration systems, in general, and property transactions, in particular, appear increasingly important.

In recent years, much attention of the international research community is paid to the comparative evaluation of land administration systems (e.g., Mansberger et al., 2000; Zevenbergen, 2002; Steudler et al., 2004;) where land administration systems are considered from a general perspective. The problems of standardisation in the cadastral domain is another intensive research topic (see the development in Oosterom and Lemmen, 2003; Lemmen et al., 2003; Lemmen et al., 2005) where structuring the cadastral information is addressed. However, the research on cadastral processes within national land administration systems is still in an embryonic phase. One of the recent attempts to compare the national property transactions is undertaken by the COST[1] Action G9 'Modelling Real Property Transactions' (Stubkjær, 2002).

This chapter is one of the outcomes of the COST Action G9. In particular, it deals with two types of property transactions (i.e. property transfer, and property formation) in three European countries, namely Denmark (DK), the United Kingdom (UK), and Belarus (BY). Historically, in the UK three jurisdictions are

[1] http://cost.cordis.lu/src/home.cfm.

distinguished, those of England and Wales (E&W), Scotland, and Northern Ireland. The article evaluates the particular jurisdiction of England and Wales. The specific institutional arrangements, from a historical perspective, concerning property transactions in all three jurisdictions of the UK are discussed in detail by Dixon-Gough and Deakin (2003).

These countries been chosen for evaluation for three reasons. Firstly, they represent three different legal families existing in the world. Denmark is a representative of the Nordic legal family, England and Wales belong to the 'Common law' family, while the modern legal system of Belarus represents the Roman-Germanic family. Thus, the chapter compares the ways in which real properties are transacted on the markets with the different legal frameworks. Secondly, in Denmark, and England and Wales, property transactions are handled in different ways but are both rationally structured and properly registered. Thus, in this respect the properties are represented for the property market. Belarus is, in turn, at the initial stage of maximising the efficiency of the property market and its problems in reorganising the market seem to be urgent. Thus, the chosen countries also vary in terms of efficiency of the property market and its institutional arrangements. Thirdly, a comparison of the property transactions in Belarus with those of other European countries might serve as a basis for developing a better solution for property transactions in the country that suit its specific legal and organisational framework.

The main goal of this chapter is to identify the general features of the national transactions, to make a comparative analyse of them, and then state their main differences. Mattsson (1997) distinguishes three general types of changes related to land, namely transfer of real property (e.g., sale, gift, inheritance, etc.), property formation (e.g., subdivision, amalgamation, etc.), and alteration of land use (through planning and environmental regulations). In particular, this chapter deals with the formalised descriptions of two types of property transactions, specifically the purchase and the subdivision of a land plot with a single-family house on it held in private ownership in Denmark, England and Wales, and Belarus. The transactions are presented as the activity diagrams in Unified Modelling Language (UML) notion with help of Microsoft Visio.

## Purchase of real property

Property purchase, besides a seller and a buyer, normally involves a real estate agent facilitating the transfer and, normally, a mortgage bank financing the purchase. Moreover, if support from a technical expert is needed, the parties hire such assistance.

To facilitate this evaluation, a purchase process in both England and Wales, and Denmark is divided into three general phases, namely pre-contracting, contracting, and registration phases[2] (Figure 10.1 and Figure 10.2, respectively). In

[2] This division is based on Ferlan et al., 2004 (unpublished).

Belarus, an additional, fourth phase, the mortgage, is recognised (Figure 10.3) since it takes place after completion, rather than during purchase.

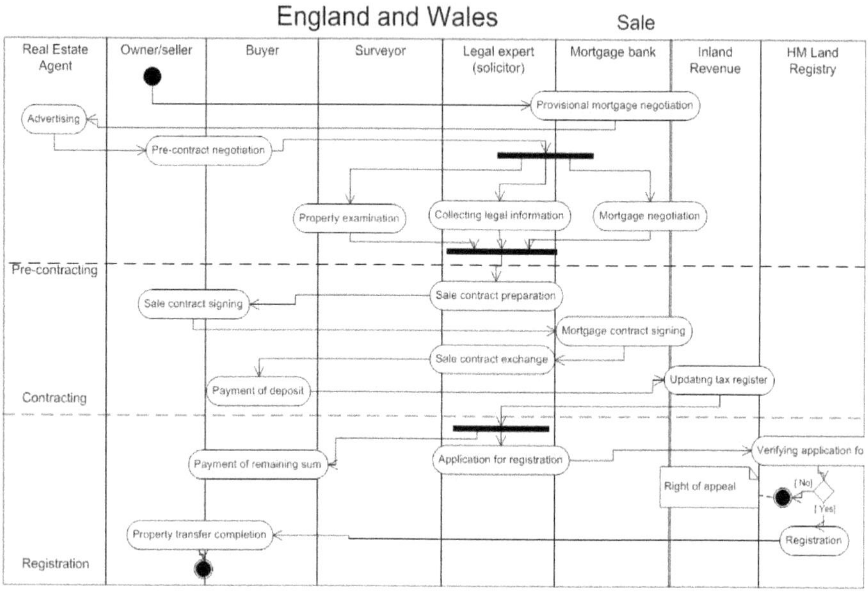


**Figure 10.1 The purchase of real property in England and Wales**

In all of the countries, *the pre-contracting* phase includes advertising by a real estate agent in newspapers, on the Internet, etc., together with pre-contract negotiations that include a pre-contract agreement implying that the parties, in principle or 'subject to contract' agree on the exchange of the property at an agreed purchase price. This agreement is legally binding if made in writing in Denmark and Belarus. In Denmark, it is effective even if made orally although a seller of a one-family house may be freed of obligations due to technical defects, by presenting to the buyer a standardized report (Danish: Tilstandsrapport), which is prepared by a licensed engineer, as well as an offer for insurance (Danish: Forsikringstilbud). In England and Wales, the agreement is only legally-binding once contracts for the sale and purchase of the property have been exchanged and either party can withdraw from the sale or purchase up until the time of the exchange of contracts.

The surveyor in England and Wales inspects the property and produces either a homebuyer report (with tick boxes) or a building survey. The former takes less time while the latter is a hand composed detailed report and includes the description of the whole property. Another activity, involving the collection of legal information on the property, is normally made by real estate agents in Denmark, or by solicitors in England and Wales.

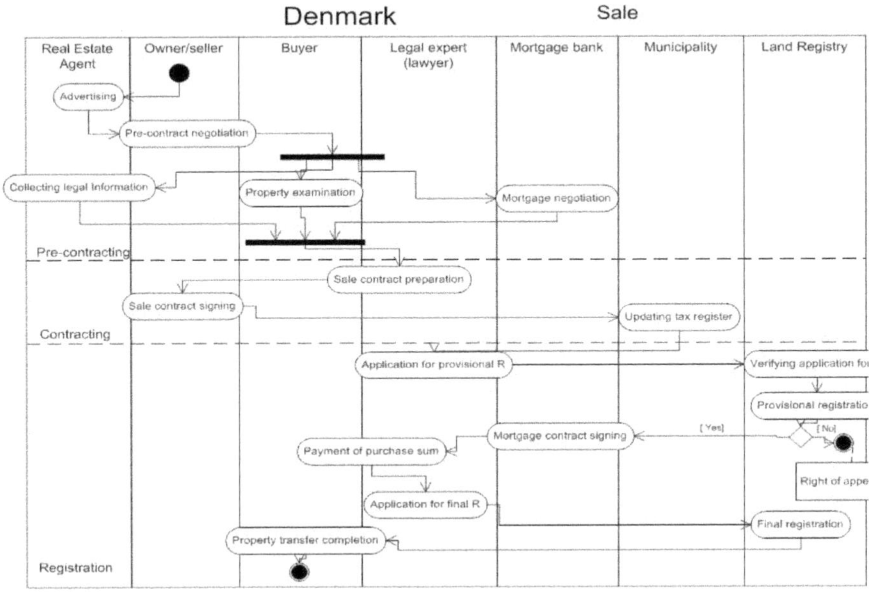


**Figure 10.2 The purchase of real property in Denmark**

The situation in Belarus differs in that a seller of property is responsible for providing all the legal information gathered from the various governmental authorities.

Mortgage negotiation is also carried out in the pre-contracting phase. A buyer would normally arrange the financing of the purchase. Before starting any negotiations in Denmark and England and Wales, a buyer contacts a bank to make a provisional agreement for a mortgage. In the UK, a buyer receives a formal mortgage offer later in this phase and accepts it by signing one copy of a mortgage contract and sending it back to the bank. In Belarus and Denmark, negotiations for purchase financing takes place simultaneously with property examination and collecting of legal information. A specific feature of Belarus is that a bank secures a loan repayment by an applicant's salary (if not sufficient, by the salaries of other family members or other – up to 9 – warrantors) and not by a mortgage against the real property (as in England and Wales and Denmark).

*The contracting phase* is very similar in all the evaluated countries and, in general, consists of preparing and signing a sale contract as well as signing a mortgage agreement with a bank (except for Denmark, where mortgage contract is signed in the later phase). In both Denmark and Belarus, a sale contract is signed by both parties simultaneously and witnessed by a legal expert (i.e. lawyer or notary respectively), while in England and Wales one additional activity – the exchange of contracts – is recognised. In England and Wales (for the detail description of a contract's content see Abbey and Richards, 1999) the contract for sale is usually signed separately by each party, in the presence of a solicitor. After signing, the signed copies of the contract are exchanged between the parties to provide a legally binding contract of sale. To protect clients, a solicitor will hold the contract until ALL contracts in the 'chain' have been signed. In the event of a breakdown in the chain, the contracts could be destroyed.

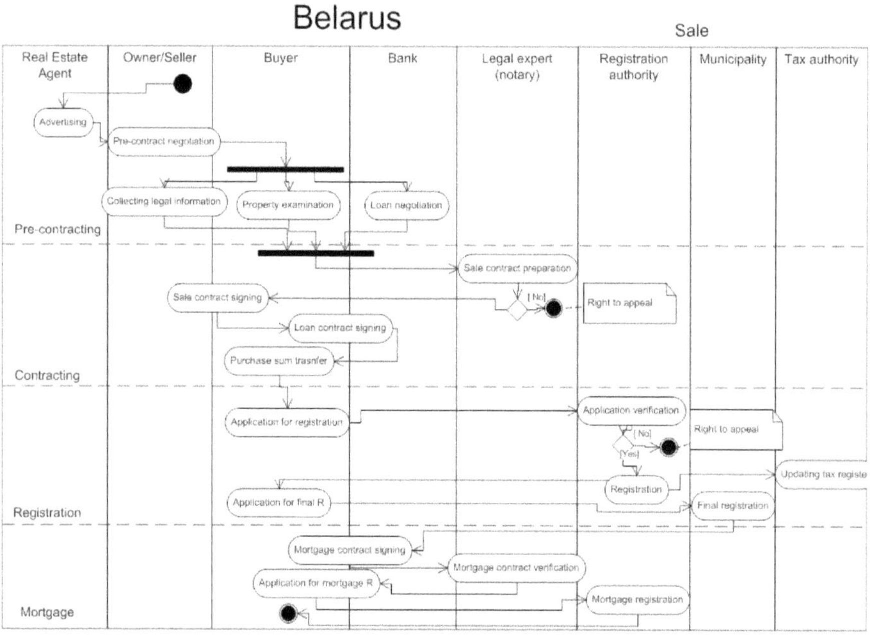


**Figure 10.3 The purchase of real property in Belarus**

Thus, in England and Wales, signing and exchange of the copies of the contract are separated in time and place, while in Denmark and Belarus these two actions are completed simultaneously.

During the signing of the contract in Belarus, a buyer normally pays the whole purchase price to a seller, while in England and Wales a deposit (normally 10 per cent of the agreed purchase price) is paid to the acting solicitor following the exchange of contracts. The remaining sum is transferred at the completion day (i.e. a moving out/moving in day). In a normal case, there is a four-week period between the exchange of contracts and completion day, although both actions can take place on the same day. The peculiarity of the Danish situation is that before contract signing, deposits are sent to real estate agent and seller's bank respectively, while complete financing of the purchase takes place in the next phase – the phase of registration.

Updating the tax register in England and Wales and Denmark is also performed after signing the sale contract. In Denmark, a municipal tax office is responsible for maintaining and updating information within municipal borders, whilst in England and Wales a solicitor applies to the Inland Revenue, using a form signed by the purchaser(s). In Belarus, the tax register is updated in the later phase, after registration.

A specific feature of England and Wales is the existence of a so-called 'registration' gap between the moment of signing the contract and registration itself. It might take up to two months for the purchase to be registered (Millgård, 2003). It can create uncertainty but is accepted and, with the current improvements in the Land Register, likely to be reduced.

In all three countries, *the registration phase* includes the main activity – registration. For England and Wales, the registration and payment of the remaining sum are the main activities, whilst in Denmark registration it is divided into the provisional and final stages, and is normally handled by the lawyer. Provisional registration is equivalent to a reservation of title, so that a seller can take back the property if a buyer fails to pay the purchase sum according to the agreement. When the Land Registry has made title reservation without conditions (Danish: Retsanmærkning) and when the purchase sum is deposited in the seller's bank, the lawyer is normally authorised to make the final registration of the deed of conveyance. The facility letter (Lånetilbud) that was the outcome of mortgage negotiations is now completed by a mortgage deed (Pantebrev), which is registered in the context of final registration. Registration is also specific in Belarus. The main distinguishing feature originates from the definition of real property in the Civil Code of the Republic of Belarus.[3] It identifies the different types of real property, namely land plots, buildings, forest, etc. A land plot and a building on it are considered as two separate real properties and, therefore, registration of land and building also differs: buildings are registered by the registration authority, while land shall be registered both by the registration authority and municipality. In other words, to become the rightful owner of a land plot with a building on it, the owner has to first register land and building in the registration authority, and secondly, to endorse the State Act on Land (issued by the registration authority) in the municipality and to additionally register it in a land book of municipality. It is only after registration and receiving all the documents certifying the ownership right on land and building, that a new owner is entitled to make a mortgage agreement with a bank.

By signing the mortgage contract, the fourth phase of the process, *the mortgage phase*, begins in Belarus. The mortgage contract shall be verified and authorised by a notary and then registered by the registration authority[4]. At present, most of the banks granting mortgages to the citizens are not actually in favour of mortgages. It is much easier for banks to secure a repayment of a loan secured by

---

[3] The Civil Code of the Republic of Belarus (1998). Minsk. http://law.by/work/EnglPortal.nsf/6e1a652fbefce34ac2256d910056d559/3e6a9e4274f90afec2256dc1002718a2?OpenDocument

[4] The State Registration Act (2002). Minsk.

income and not by real properties. It can be explained by the strict housing legislation in Belarus stating that a family with dependent(s), in a case of failure to pay the loan back, cannot be moved out from the residential property without another place being provided for them to live. Thus, the existing property legislation in Belarus hampers the development of the mortgage market, in particular, and the capitalisation of the properties on the market, in general.

## Subdivision of real property

For the purpose of the comparative evaluation, property subdivision processes in England and Wales, Denmark, and Belarus are divided into three general phases. This section discusses the subdivision of a land plot with a single-family house.

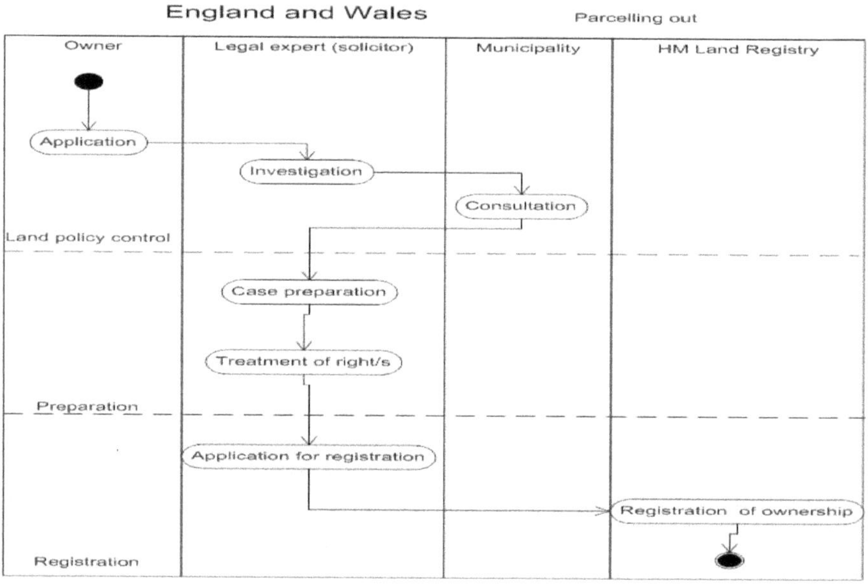


**Figure 10.4 Subdivision of real property in England and Wales**

In this example, the 'old' land plot retains the residential house and the new one has no building(s) on it. Land privatisation, as a specific case of property subdivision in Belarus, is also addressed.

Property subdivision (or parcelling out in England and Wales) means that new property unit(s) with unique cadastral number(s) (reference number(s) in England and Wales) are created and registered. These are land policy control, preparation, and registration.[5] This division is in accordance with the specific functions (i.e., purposes) of the activities presented. Land policy control in Denmark, and England and Wales includes case investigation when an initial judgement of subdivision is performed either by a solicitor (England and Wales) or a surveyor (in Denmark). For Denmark and Belarus, obtaining approval from local government and/or the different governmental authorities supplement the case investigation. Preparation includes the activities both in the field (e.g., surveying measurements) and in the office (e.g., investigation and collection of information) and, finally, registration consists of the official recognition of the subdivided properties.

Land policy control in England and Wales (Figure 10.4) is relatively minor and consists of an investigation, normally by a solicitor, and consultations with local government (if required). Before visiting a solicitor, the owner measures the new boundaries and, thus, makes a decision concerning the location of new boundaries. Furthermore, the owner may also impose additional restrictions on the new plot, whilst any restrictions relating to 'old' parcel will automatically be transferred to the new one. In this phase, the owner and the solicitor are the key actors. No approval is required from central government not, in most cases, from regional or local government.

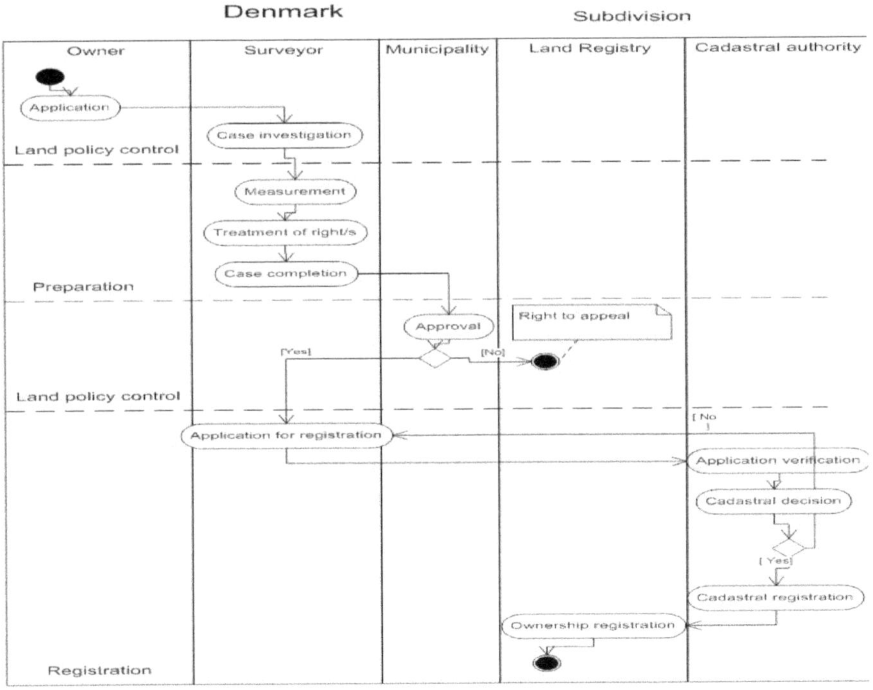


**Figure 10.5 Subdivision of real property in Denmark**

[5] This division is based on Ferlan et al., 2004 (unpublished)

Depending upon the local planning policy and the nature of the boundary that will be erected between the new and old parcel, no approval (only consultation) is required. If the new parcel is to be used for building purposes, planning approval must be sought for the construction of the new building but not for the purpose of subdivision *per se*.

In Denmark, land policy control is implemented through two stages, namely before and after the preparation phase (Figure 10.5). Firstly, a licensed surveyor, who is legally and technically responsible for property subdivision, makes an initial judgement of the case by investigating the relevant legislation (i.e. agricultural, environmental, land use and building act, road visible features (e.g., hedges, stone walls, etc.) is clearly stated (Dixon-Gough and Deakin, 2003: 86). A solicitor draws the boundary lines on an existing map, and composes a verbal description as a separate document, which will also the transfer of rights, including the creation of easements and interests (e.g., the mortgage provider) from the 'old' to 'new' parcel. A solicitor also composes a title deed for a new plot and makes amendments into the 'old' one. However, this formalised procedure will only take place if the 'new' parcel is to be sold, or transferred to a different owner. If the parcel is subdivided yet stays within the original ownership, there is no need for any of the above actions.

In Denmark, a surveyor measures the newly subdivided plot, calculates its area, and adjusts the position of the existing boundaries on the cadastral map according to the measurements made. Moreover, the surveyor investigates whether the exiting easement rights shall belong to the original or to the new property unit or to both. The surveyor can also create new easements if needed with the consent of the owner. Thus, a surveyor in Denmark carries out any treatment of rights, whilst in England and Wales a solicitor deals with it (municipal building statutes, as well as ancient monuments). If the subdivision is for building purposes, the surveyor also checks the building regulations concerning the purpose, position and the appearance of the building. The second stage begins when a licensed surveyor applies to municipality and other relevant authorities for approval of subdivision.

The municipality and other local authorities (e.g. environmental, agricultural, road authorities, etc.) approve the case according to the existing regulations. In the case of a negative response, the owner is entitled to appeal against the decision taken. These two stages of land policy control are separated by the preparation phase (including measurement of boundaries). To a great extent, this process is very similar to that of the England and Wales in those instances when the subdivided land is to be used for residential and industrial building purposes.

Land policy control in Belarus seems to be the most extensive of the evaluated countries (Figure 10.6). The key actors are a municipality and a surveyor. The latter investigates and prepares the case for two sets of municipal approval; a preliminary case approval, and an approval of new plot's location. An applicant (i.e., a new owner) is responsible for making a contract with a surveying organisation.

The preparation phase in Denmark and Belarus includes preparation and completion of property subdivision, treatment of rights, and measurements. In

England and Wales surveying measurements are not normally made, since cadastre, in the commonly accepted sense, does not exist and, thus, land plots are not geometrically defined on the surface. Instead, the concept of general boundaries is widely applied in England and Wales, which means that the position of the boundary is not exactly defined but the location of the parcel in respect to visible features (e.g., hedges, stone walls, etc.) is clearly stated (Dixon-Gough and Deakin, 2003: 86). procedure will only take place if the 'new' parcel is to be sold, or transferred to a different owner. If the parcel is subdivided yet stays within the original ownership, there is no need for any of the above actions.

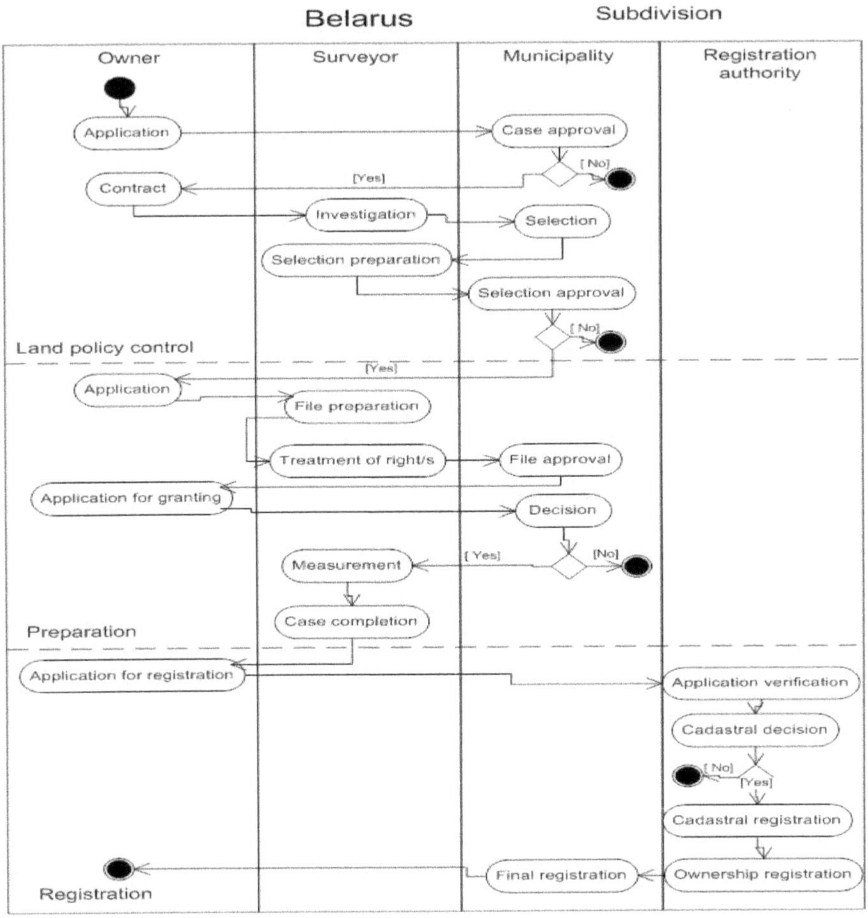


**Figure 10.6 Subdivision of real property in Belarus**

A solicitor draws the boundary lines on an existing map, and composes a verbal description as a separate document, which will also the transfer of rights, including the creation of easements and interests (e.g., the mortgage provider) from the 'old' to 'new' parcel. A solicitor also composes a title deed for a new plot and makes amendments into the 'old' one. In Denmark, a surveyor measures the newly subdivided plot, calculates its area, and adjusts the position of the existing boundaries on the cadastral map according to the measurements made. Moreover, the surveyor investigates whether the exiting easement rights shall belong to the original or to the new property unit or to both. The surveyor can also create new easements if needed with the consent of the owner. Thus, a surveyor in Denmark carries out any treatment of rights, whilst in England and Wales a solicitor deals with it.

In Belarus, the preparation phase is also the most extensive one among three evaluated countries and includes such stages as case approval (together with a case file) by the municipality, together with a municipal decision concerning the withdrawal of land from the 'old' owner and granting it to the new one. The rights over the subdivided plots are handled by a surveyor but approved by the municipality. Furthermore, measurements in Belarus are carried out only after a decision by the municipality whilst in Denmark they are made before municipal approval. Unlike in Denmark and England and Wales, the municipality in Belarus is involved in this phase. In particular, the municipality approves the case (together with a case file) whilst in Denmark and England and Wales, local government is only involved in the earlier stage (i.e., the land policy control phase). One more difference between Belarus and two other countries is that an owner in Belarus also takes part in this phase by making application to both a surveyor and the municipality.

The registration phase is the most important since the newly established property rights are to be registered and, thereby, can be dealt with separately in the future. It is organised in accordance with the national institutional arrangements for property registration of the respective countries. In Denmark, two independent governmental authorities maintain the cadastre and land register, and are responsible to different Ministries. This phase in Denmark begins with an application submitted by a surveyor to the cadastral authority. The application (i.e., all the documents submitted by a surveyor) is checked by a cadastral registrar. If no inconsistencies are found, the decision about registration is taken, and the cadastral authority informs the Land Registry, the municipality, and the surveyor. The municipality updates the Real Property Data System,[6] and since the latter is automatically connected to the Tax register, the tax authority is informed about the changes made as well.

In Belarus, the cadastre and land register are unified and maintained by a single registration authority and, thus, cadastral and ownership registration is carried out simultaneously. The distinctive activity of the registration phase in Belarus is the final registration of land ownership by the municipality. It means

[6] http://www.fsk.dk/fsk/publ/1996/it96-uk/inde0014.htm, initiative 5.4.

that to become the rightful owner of a subdivided plot, the owner must possess the State Act on Land endorsed by the Head of the municipality and recorded in the municipal land book.

In England and Wales the registration phase is the smallest among the countries. It only includes registration in Her Majesty's (HM) Land Registry. Since the concept of general boundaries is historically accepted in England and Wales, it, in turn, led to the absence of cadastre as a broadly recognised term. Therefore, there are no such activities as cadastral decision and cadastral registration in England and Wales. On the one hand, it speeds up property subdivision but, conversely, it can introduce problems leading to boundary disputes. Another distinguishing feature of this phase in England and Wales is absence of application verification. In other words, HM Land Registry just awards a title number to a new plot, enters records/changes in three parts of the register (i.e. the Property Register, the Proprietorship Register and the Charge register in the case of mortgage or other financial burdens on the property) without verifying the submitted documents, whilst in Denmark and Belarus application verification is an obligatory activity.

## Conclusions

### *Purchase of real property*

A comparative evaluation of the key actors involved in purchase has shown that Denmark and England and Wales have, in general, a more user-friendly process in comparison with Belarus. In the latter, a seller and buyer are responsible for most of the activities, while in England and Wales a solicitor facilitates the process. In Denmark, responsibilities are rather equally distributed among the parties, although a legal expert plays an important role. According to the evaluation, Denmark is considered as the country where the role of parties is valued relative to that of experts and registration services.

The contracting phase is the one that discriminates the most among the evaluated countries. It is distinguished by the time and place for signing and exchange of the contracts: in England and Wales the parties are normally split, whilst in Belarus and Denmark the contracts are signed and exchanged simultaneously. The financing of the purchase is also arranged in this phase, although, the timing of the payment of the purchase sum varies. It would seem that a buyer in Denmark is more protected against a seller's fraud than in both Belarus and England and Wales. Furthermore, in both England and Wales and Denmark, financing is normally arranged through a mortgage contract with a bank or mortgage institute, whilst in Belarus the purchased property might be mortgaged only after registration. Moreover, banks in Belarus prefer to secure loans through the incomes of the applicants rather than through real properties and thus, real properties are not capitalised on the market. The reason why banks are not in favour of mortgages is the virtual impossibility to seize the properties in the case of payment failure because of the existing housing legislation in Belarus. Thus, the need for amendments in the housing legislation of Belarus is clearly stated.

Updating the tax register also takes place in different phases of the process. In Belarus, it is updated after registration, but in Denmark and England and Wales it is immediately after the signing of the sale contract. The reasons for this might be not only the differently organised national institutional frameworks but also a low level of IT services in Belarus on the governmental level.

Belarus differs from other countries also in a way a land plot is registered. Whilst in Denmark and England and Wales, ownership registration takes place in the Land Registry, in Belarus the rights on the land must additionally be registered in the municipality. That in turn lengthens the whole process and makes it more burdensome for the parties.

Generally speaking, registration in Denmark and Belarus could be identified as the most important activity. However, in England and Wales the exchange of contracts is seen as the most important activity since registration is still significantly delayed from the transfer completion.

In summary, the purchase process in Belarus seems to be more complicated and burdensome for the parties in comparison with those of Denmark and England and Wales. Moreover, it seems that the mortgaging of property in Belarus is almost a superfluous activity for the banks, since they cannot take the mortgaged property back under specific circumstances in a case of payment failure. The most secured process from seller and buyer's point of view is seen in Denmark where the buyer is protected from fraud by provisional registration. The biggest technical drawback of the process is found in England and Wales in the existence of a 'registration' gap between the signing the sale contract and registration itself. The period might be up to two months, during which the new owner is not protected against a seller's potential fraud. This is largely a technicality since the forms, signed by the purchaser and seller, are lodged with the Land Registry, whilst the contract for sale has been witnessed and exchanged. As such, the likelihood of fraud is unlikely.

Thus, the evaluation of property purchase has clearly identified the differences in institutional arrangements of the countries as well as their shortcomings and drawbacks.

### *Subdivision of real property*

The most distinctive activity of property subdivision among the countries is the surveying measurements. In England and Wales, no accurate measurements are made in the field although the boundary may be marked at either end with wooden pegs, whilst in Denmark and Belarus they consist of geodetic calculations and the physical demarcation of boundaries. However, the time schedule for measurements varies between Denmark and Belarus. The subdivided plots are normally demarcated before the municipal decision in Denmark but after it in Belarus. This raises the question of whether it is rational for the owner in Denmark that the measurements are made before approval? In some cases, the municipality can reject the subdivision and the owner will have to bear the cost. To avoid such a situation, the surveyor postpones demarcation in the field in those cases where the decision of local authorities is hard to predict. Since measurements in Belarus are performed only after the final decision of the municipality, refusal in the beginning of the process will not lead to heavy expenses for an owner.

Moreover, cadastre does not exist *per se* in England and Wales and, therefore, there no need for cadastral procedures. It simplifies subdivision in England and Wales in comparison with that of Denmark and Belarus. Belarus is, as mentioned above, distinguished by double registration of rights on land, both in the registration authority and the municipality. It is still unclear why the municipality duplicates the registration thereby increasing costs for the owner?

To better understand the national institutional frameworks, the key actor of subdivision process in all the three countries have also been identified: a solicitor carries out the investigation and applies for ownership registration in England and Wales, whilst a surveyor takes technical and legal responsibilities in Denmark. In Belarus, a surveyor acts only as a technical assistant whilst the municipality has all the legal power. In addition, the level of formalities in the countries differs greatly. In Denmark and England and Wales, subdivision is informally initiated, whilst in Belarus there is a strict, formal procedure. Land policy control in Belarus and Denmark also requires formal approval from the municipality that, in turn, increases the time for subdivision. In England and Wales, the municipality is not concerned and thus this phase is less formalised in comparison with two other countries. It is recognised that the more formal process leads to the higher transaction cost in term of time and money.

Furthermore, in Belarus, ownership registration is mandatory, whilst in Denmark and England and Wales, subdivision as such does not change the owner status. Thus, the Land Registry records that the owner now owns two or more property units. The selling of one of such units is another case.

To conclude, property subdivision in England and Wales appears to be the user-friendliest and less formalised, and includes fewer activities in comparison with those of Denmark and Belarus. Property subdivision in Belarus seems to be burdensome and time consuming for owners with many approvals required. Moreover, an owner is responsible for the process, through making a number of applications to the municipality and contracts with a surveyor. Thus, it is obvious that property subdivision in Belarus needs simplification, i.e. to be approached from the users' perspective through decreasing the number of approvals and applications. Will it be reasonable to leave just one municipal approval in the beginning of the process? What are the reasons for keeping the final registration by municipality? What is it meant for? What will happen if it is abolished? Surely, these questions need to be further investigated.

## References

Abbey, R.M. and Richards, M.B., 1999. *Conveyancing*, Blackstone Press Ltd, London.

Dixon-Gough, R.W. and Deakin, M. (2003). Property transactions in the UK: a situation of institutional stability or technical change? In: Stuckenschmidt, H., Stubkjaer, E. and Schlieder, C. (eds.), *The Ontology and Modelling of Real Estate Transactions*, Ashgate, Aldershot.

Lemmen, C, Molen, P., Oosterom, P., Ploeger, H., Quak, W., Stoter, J., and Zevenbergen, J., 2003. A modular standard for the cadastral domain, presented to the *3rd International Symposium on Digital Earth*, Brno, The Czech Republic. Available at: http://www.gdmc.nl/publications/2005/Modular_cadastral_standard.pdf.

Lemmen, C., Oosterom, P., Zevenbergen, J., Quak, W. and Molen, P., 2005. Further progress in the development of the core cadastral domain model, presented at the *FIG Working Week 2005 and 8th International Conference on the Global Spatial Data Infrastructure: From Pharaohs to Geoinformatics*, Cairo, Egypt. Available at http://www.fig.net/pub/cairo/index.htm [accessed on 25 June 2005]

Mansberger, R., Dixon-Gough, R.W. and Seher, W., 2000. A comparative evaluation of land registration and agrarian reform in Austria and Great Britain. In: Dixon-Gough, R.W. and Mansberger, R. (eds.), *Transactions in International Land Management*, Vol. 1, Ashgate, Aldershot.

Mattsson H., 1997. The need for dynamism in land, *Land Law in Action*, Swedish Ministry for Foreign Affairs and Kungl, Stockholm, Tekniska Högskolan.

Millgård, O., 2003. *Conveyancing of Land, and the Role of Registration, in Germany, the Nordic Countries, England and Wales*, Report 4:94, KTH-Infrastructure, Real Estate Planning and Land Law, Stockholm.

Oosterom, P. and Lemmen C., 2003. Towards a standard for the cadastral domain: proposal to establish a core cadastral data model, presented at the *Workshop Cadastral Data Modelling*, Enschede, the Netherlands. Available at: http://www.oicrf.org/ [accessed on 18 June 2005]

Steudler, D., Rajabifard, A. and Williamson, I.P., 2004. Evaluation of land administration systems. *Land Use Policy*, **21**, 371-380.

Stubkjaer, E., 2002. Modelling real property transactions, *Proceedings of the XXII FIG International Congress*. Washington, DC. Available at: http://www.fig.net/pub/fig_2002/Js14/JS14_stubkjaer.pdf [accessed on 1 July 2005].

Zevenbegren, J., 2002. *Systems of Land Registration – Aspects and Effects*. PhD thesis. Publication on Geodesy 51, Delft, the Netherlands.

# Chapter 11

# The Role of Land Consolidation and Land Readjustment in Modern Society

Robert Dixon-Gough

## Introduction

Rognes and Sky (1998) describe land consolidation in the Norwegian context, the official name of which is '*jordskifte*', which is normally translated in English as 'land consolidation'. The more precise translation of *jordskifte* in English would be, 'the reallocation of holdings by pooling and redistribution'. This term expresses the fact that during a land consolidation process many land parcels from a defined number of land holdings are combined, and from the same area of land, the holdings emerge but in new physical and legally recognisable shapes. Land consolidation is normally carried out for all the holdings within a very specific geographically definable area. The size and extent of land consolidation varies from minor readjustments of boundaries between two adjacent holdings, to the complete rearrangement of hundreds of holdings with planning and investment in a new rural or urban infrastructure. At a fundamental level, the land consolidation and land readjustment process is intended to restructure outdated or unsatisfactory land ownership patterns.

Riddell and Rembold (2002) consider that one of the main justifications for land consolidation, apart from land fragmentation through inheritance, is that land parcels in many parts of the World, and particularly those in much of Europe, are unsuited in both size and distribution to the kind of rural economy that is needed for farming in the Europe of the early twenty-first century.

The reasons for this are partly the result of recent history. In the case of many states of the former Soviet Union, the processes of land restitution that followed 1989 were a combination of politics, social justice, and some degree of economic consideration (van Dijk, 2003a, 2003b; 2004). Since many of the processes of restitution were based upon the old cadastres of the 1930s, or in some cases the 1910s, the size, location and valuation have more relationship to the economic situation during those periods and the immediate post war period than it did for the current situation.

One of the products of such a process is that the situation of restituted property completely disregards more than 50 years of centrally planned infrastructure development (communications, irrigation, drainage, etc.). In addition to the problem of the restituted property, the social structure of the rural areas has changed significantly since 1989. The proportion of people whose prime occupation is directly related to farming has declined, whilst the urban areas are now the 'dynamic engines', ensuring that all future rural developments will have an urban bias since the majority of the population now live in urban areas.

To counter this problem of rapid urban growth, an interdisciplinary group of the FAO was assigned to develop a new programme to provide assistance to rural communities. The basic concept was to combine classic land consolidation techniques with participatory approaches through spatial data technology (Riddell and Rembold, 2002). This approach is based upon the following principles:

- land consolidation must be participatory, democratic and driven by the community;
- the focus must be upon rural livelihoods rather than upon the production of food staples;
- the end result will be community renewal, that is, sustainable economic and political development of the whole community; and
- it is founded on the principle of assisting the community define new uses of its resources and then reorganise its spatial components (parcels) accordingly.

Issues relating to land-tenure affect virtually every decision in countries that have an emphasis on the rural economy (Fearnside, 2001). These will range from the investments in labour and capital by landowners of all sizes, to the migrations of populations from rural to urban areas, the formation and action of social movements, to the initiation of governmental, regional and international programmes. Land use and its influence upon the environment is a direct outcome of such divisions.

Furthermore, changes in land tenure procedures is a central element in the means of directing land tenure and land consolidation towards a development process leading to greater awareness and practice in sustainability, socially beneficial, and environmentally sound land management procedures.

This can be emphasised in the concept of integrated agricultural production (IAP), which lies somewhere between conventional and production agriculture. Whilst conventional agriculture may be considered to be the evolution of production methods aimed at feeding an increasing population and establishing national and regional self-sufficiency, changing trends in co-operation, falling populations and changing forms of consumption means that such techniques are now no longer a necessity. The basic assumption of the IAP system attempts to address these problems whilst maintaining relatively high levels of agricultural production and is ideally directed towards specific regions rather than upon individual farms. IAP addresses the physical structure as a means of improving the ecological balance within that area. Wilkowski and Sobolewska (1998) have identified two groups of activities that are integral to IAP and which relate specifically to the ecological infrastructure. These are activities relating to:

- crop production, such as belts of trees between trees, buffer belts (green highways) along roads or streams, inter-field strips of natural vegetation (areas of ecological compensation) all of which are aimed at developing a *natural infrastructure*;
- the management of farms and rural areas such as the treatment and disposal of human, animal and farm waste, and the development of suitable storage and disposal of pesticides, all of which are aimed at the *technical infrastructure*.

Both activities are concerned with the transformation of the physical structure of rural areas and must be implemented by experts in the management of rural areas, typically those who implement land consolidation projects. One of the key elements of land consolidation should be the consideration of the requirements of IAP and to develop the relationships and mechanisms between agricultural production, the natural ecology and the retention and possible restoration of key landscape elements.

In Poland, for example, land consolidation activities address the philosophy of IAP in the development of rural areas. This consists of considering:

- the quality of life and meeting the aspirations of rural societies;
- spatial management and the preservation of the environment aimed at developing an harmonious relationship between the environment, ecosystems and the optimum management of their related elements;
- considerations of the multi-functional characteristic of rural areas, transforming the conventional characteristics (primarily related to the optimum agricultural production) into the provision of an integrated rural society with an increased variety of professional and job opportunities;
- meeting the requirements of regional policies of the EU through access to regional funds. This will, however, introduce new problems as the result of increased competition.

## Policy issues of land reform and rural land consolidation

Williamson and Ting (2001) comment that regional, global, and political drivers such as sustainable development, land reform, urbanisation, economic reform, and technology are constantly changing the ways in which land is viewed and perceived. These changes require new land administration infrastructures and tools and, as a result, existing land administration, land management and cadastral systems are being re-engineered.

Land administration processes, including cadastre and land consolidation, can no longer rely upon manual processes or traditional structures that had been developed to support former economic systems. To utilise systems that support individual processes and in which data and processes were maintained separately, such as land valuation and land titling, is no longer sustainable. They are being replaced by multipurpose land data banks in which information relating to resources, planning, land use, land value, land titles (including common property rights) can be integrated for a range of land administration and business purposes. Such land databanks (or multi-purpose cadastres) have a much broader and integrated vision than has been experienced in the past. The components of land registration, cadastral surveying, large-scale topographic mapping, soil fertility, planning, and land valuation, together with their inter-relationships in the sphere of land markets, should be considered as an integrated system in which the common objective is sustainable development.

One of the main advantages of such an integrated approach is that all rights, restrictions, and responsibilities (including both two- and three-dimensional interests) relating to land must be considered in designing and implementing such a land administration system together with its integrated land databank. Such a system would be of immense value to the processes of land consolidation.

Despite certain commonalties, land fragmentation patterns differ from country to country and it is possible to distinguish between fragmentation of ownership and fragmentation of land use (Riddell and Rembold, 2002). The Czech Republic and Hungary managed to control extreme forms of land use fragmentation for the time being due to fact that co-operative farms were not dismantled during land reform and the current legislation allows lease agreements to increase farm size. Nevertheless, transaction costs are high given the sheer number of short-term lease contracts and legal and juridical restrictions regarding freehold arrangements, which still hamper farm enlargement. Such is the case of Hungary, where co-operatives, according to the current legislation, are not allowed to purchase land.

In this context, it still remains uncertain whether the cooperative-type of farms will survive in the long run especially in view of EU common agriculture policy and open market conditions. If a significant number of these farms fail to be competitive, the problem of fragmentation could resume on a significant scale. This is clearly shown in the case of Romania where fragmentation of ownership coincides largely with the fragmentation of use of the land since most cooperatives were dismantled at the outset of transition and this farm type virtually disappeared in 1992. At the moment the average size of private individual farms, which account for 62 per cent of agriculture land is 2.3 hectares spread over 6-10 parcels. Estimates for Bulgaria indicate, that once the land reform is finalised, more than 2.6 million private farmland titles, divided among 12 million parcels with an average size of 0.4-0.5 ha each, will be issued. The average size of the holding is approximately 2 ha (Riddell and Rembold, 2000).

Thus, after a decade of farm privatisation and restructuring, the organisation and structure of farming in most of Central and Eastern European (CEE) and the Commonwealth of Independent States (CIS) has changed dramatically (Graefen, 2002). Today a large portion of farming is carried out on private land in individual farms, with distinct differences between the two regions CEE and CIS.

Regardless of the dynamics of change, efficiency, and productivity in terms of land, capital and labour still remains unsatisfactory, and is even declined in a number of countries throughout the 1990s and beyond. Land fragmentation has increased with few exceptions throughout the whole region, putting an additional burden on farm management. Implications for the rural areas tend to be dramatic in terms of an aging population, migration to urban centres, increasing rural unemployment, and strong economic differentiation among the regions and sub-regions of individual countries.

A decreasing population in all CEEC countries, with the exception of Albania and Bosnia, has hit rural areas very hard. For example, the overall population of Bulgaria decreased from 1991 to 1999 by 405,500, or 4.71 per cent. The tendency is continuing and grew stronger during the latter part of the 1990s and was aggravated in the marginal rural regions. 'Ghost' and completely deserted villages are common proof of this development. A large portion of all arable land is effectively unused, e.g. 25 per cent in Estonia. Thus, rural land markets are depressed and restricted to favourable regions in CEEC.

*Land reform as part of the agricultural transformation in the CEE and CIS*

The privatisation of farms, particularly in the CEE countries, ranges from levels of close to 100 per cent of arable land in Albania, Latvia and Slovenia, to about 40 per cent in the Czech Republic. The average level of privatised and restituted land ranges throughout the CEE region is between 50 to 80 per cent of all arable lands. Associated with privatisation and restitution are the problems of land fragmentation, which have emerged and are currently one of the most acute constraints upon agricultural production (FAO, 2003). This is in particular true in a situation, which is characterised by:

- *Continued land insecurity.* Title registration and access to titling is often still not yet accomplished in many of the States in Central and Eastern Europe and might be established for a limited percentage of arable land (for example, approximately 50 per cent in the case of Bulgaria). Secure tenure rights are considered a pre-condition for economic development for several important reasons:
    - to encourage investment in agriculture by assuring landholders that they will be able to reap their returns on investment;
    - to increase the use of credit for the same reason and to enhance the collateral value of land;
    - to enable the transfer of land from less to more efficient uses through land transactions, by enhancing the certainty of contracts and lowering enforcement costs; and
    - to contribute towards an increased political stability by providing small farmers with a more significant stake in society;
- *Land markets do not function efficiently.* Deeds markets are in their infancy almost everywhere and land sales, even in those cases where registration has been completed, is negligible;
- *Land leasing is generally still uncommon and insecure for the both the lessee and lessor, despite the apparent increase in the number of contracts.*

The leasing of individual farm land ranges from between 2 per cent of farms in Georgia to 17 per cent in Poland and is often restricted to short-term contracts (1-2 years); *Difficulties of marketing and market access, machinery and credit supplies, and immobile labour markets.* These constitute further constraints to the enlargement of farm sizes into viable economic entities.

Even in countries like the Czech Republic, farm restructuring is considered necessary in order to continue, not only for corporate farms but all types of farm. A large numbers of individual farms make a loss and have very low levels of productivity. During the last 10 years the proportion of part-time farmers has remained quite stable at between 26 and 28 per cent.

An example of one of the more advanced transitional countries is Hungary for which the present level of reform with regard to land tenure policy is based upon the following principles:

- the completion of privatisation for agricultural lands as soon as possible;
- the intention to facilitate the consolidation of fragmented parcels by supporting voluntary exchange of lands, or through purchasing lands to be subsequently merged using the present framework of agricultural subsidies;
- the adoption of the draft law on Land Consolidation by Parliament as soon as possible;
- the adoption of the draft law on National Land Fund.

It is anticipated that tools such as the new legislation for land consolidation, which regulates the share of costs among interested parties and includes the provision of long-term low interest credit rates by the state, together with the establishment of the National Land Fund (based on state-owned areas) will ensure:

- a more adequate parcel size, as is required for viable and competitive family farming that will contribute to sustainable agriculture and rural development;
- the strengthening of the land market in general and provision of a stable market based on reasonable transaction prices;
- promotion of the State's proposal to change land use patterns especially in areas endangered by frequent floods or soil erosion.

The two-year budget adopted by the Hungarian Parliament in 2000 provides financial resources for the implementation of the land tenure policy programme, with the expectation that draft laws will be adopted in 2001.

Drees (2002) considers that an important aim of European policy should be to align living conditions between the different countries, and also between urban and rural areas. Approximately 80 per cent of the land area of Europe is rural and this has been subjected to many changes over the last decade. These changes may be experienced across Europe and include factors such as (Drees, 2002):

- many people are giving up farming and leaving the rural areas for an easier life in urban areas;
- young people, particularly the better educated and more ambitious, are leaving their home villages for professional and technical education and are not returning;
- farming is increasingly being conducted by an increasingly older population;
- rural areas are becoming de-populated and the problems of urban areas is that of over population;
- rural settlements are increasingly becoming the preserve of middle class holiday makers and weekenders.

To reduce this impact, the living conditions and aspirations of rural populations must be increased by developing and strengthening rural areas and making them more appealing and attractive places to live. This means not only improving the agrarian structure but also providing support for the non-agrarian sector to a full range of employment, improving the rural infrastructure, the environment and providing and suitable range of affordable housing.

In the parts of the former East Germany, the restitution and reapportionment of land has resulted in a high degree of fragmented parcels. In addition, land is also needed for the development of rural infrastructures, including the provision of new motorways, housing, or the development of new industries. It is estimated that such projects consume about 150 ha of land per day. This results in land dispute.

One way of overcoming these problems is to resort to land consolidation. Through the process of exchange and re-allocation, the property of state-owned co-operatives can be privatised, new farmers can be encouraged to start their own private farms, and land use problems resolved. The process can also deliver support for environmental and infrastructural projects such as housing, industrialisation, and tourism.

## Organisation of the Cadastre for land consolidation and land databanks

Riddell and Rembold (2002) have indicated that the future of land consolidation is dependent upon an ability to develop a combined land databank that will proved a single conceptual package encompassing the needs, capabilities, and aspirations of the rural populations together with the knowledge of agro-ecological planning, farm-gate level agricultural economics, and sound sustainable environmental planning. This is a very idealistic concept that will, unfortunately be beyond the financial capabilities of most states.

Spatial databases are, however, fundamental tools for decision making in the field of land use and development. For them to work in a satisfactory manner there must be active co-operation in the processing of the spatial data between both the private and public sectors (Mansberger and Muggenhuber, 1999). Furthermore, land databases must be user-friendly and provide fast and provide easy access to multiple data bases. Whilst governmental and regional policies might change the need for data remains constant.

## The role of education in rural development and land consolidation

Márkus (2001) has observed that the speed of change in the fields of data acquisition and information technology, coupled to the increasing needs for land and spatial information, and to lifelong learning provides challenges for the education and re-skilling of all professions. This is particularly so in the case of land administration and land management, both of which are now coming under increasing pressure to provide land related information and processes. One of the key features in this scenario is the need for continuing professional development (CPD).

The move towards the 'information society' cannot be reversed and in the land administration/land management sector, this is especially true. Information concerning land can be used for a range of disciplines ranging though the various processes of land management to the creation of new service markets. An improved use and application of land information can, through the potential of the information and communication technologies both:

- facilitate the provision of services by private rather than the public sector, which will include a new partnership between both the private and public sector that, in the case of training might also include the university sector; and
- speed up the processes of both administration and decision-making procedures.

One of the most efficient ways of achieving this is to adopt a system of education and training designed to implement these measures through the reorganisation of educational resources in partnership with the needs of the employers.

It is important, however, these 'new' skills are underpinned by a thorough knowledge and understanding of the basic principles and techniques of the professions

## Training and education in Hungary

The staff development strategy of the Hungarian Land Offices has been based upon the development of an education and professional development programme for those involved in the various aspects of land administration. One of the critical weaknesses encountered in Hungary (and in many other CEE countries) has been in addressing the EU criteria of a functioning market economy together with that of 'institution building'. Márkus (1998) at the UN Meeting of Officials in Land Administration (MOLA) identified some those weaknesses as a:

> Lack of education in the management, legal, economic, human, and ethical aspects of land administration, a general lack of user-orientated aspects of land administration, lack of continuity in education from universities to professions, and appropriate linkage between the two.

It is generally recognised that human resources are one of the most important elements of service delivery in all aspects of land administration and management. In many countries, those resources have often been neglected through the speed of change in information technology, especially when placed in the context of transitional changes in land ownership and the measurement of those changes. In such cases, the need for education and training is essential. Furthermore, the system should be responsive to the needs of those human resources and, with this in mind, systems of open and distance learning have been successfully applied.

In the case of Hungary, this programme of education and training has to be placed in the context of their land database and the modernisation of land administration processes. During the last decade, this has been extensively modernised and is now maintained by the network of land offices. The land database currently contains approximately 9 million land parcels and/or real estate and the topographic and topological data from more than 55,000 cadastral maps. It is, therefore, essential that the data is handled in an efficient manner by properly trained staff to reinforce consistency, common identity, a shared corporate culture, common actions, clear responsibilities, co-ordination, and the dissemination of good practice.

The main training and education programmes used in Hungary have been Márkus (2001):

- *Open Learning for Land Offices (OLLA) – the TEMPUS project* directed towards the Regional and District Land Offices in which the property records (including the large-scale cadastral maps) are maintained and up-dated. The project was formed through the co-operation of the University of Veszprém, the University of East London, and the Katholieke Universitaet of Leuven and had the aim of developing the professional skills to equip staff to work effectively in land administration activities for both the state and private sectors. This project was developed as a series of study modules relating to the fundamental issues of professional work, management, the use of data, and information technology. Although originally written in Hungarian, the core of the programme was developed to meet the demands of other systems in Eastern and Central Europe;
- *Distance Learning in GIS – a PHARE project.* This was structured to form an open and distance learning course with the aim of reaching the standards of a European professional degree. The intended target group consisted of professionals in geoinformation and related sectors, including national mapping organisations, and application orientated organisations – including cadastre, local authorities, utility companies, and private mapping companies. In conjunction with the ITC, seven modules have been prepared;

- *UNIPHORM – a PHARE multi-country project in distance learning.* This was developed for the professional development in GIS of technical staff and is offered through web-based distance learning. This programme is determined by three factors; the adoption of GIS ideas by management, the capacity for rapid change in organisations, and improving the skills level of engineers and technicians. The programme is targeted towards public administrations, regional planning authorities, land managers, and GIS companies. In particular, the programme will provide education and training in IT development for practising engineers and surveyors; adult professional needing retraining and skills updating; and university graduates requiring specific skills to enter the labour market;
- *Staff Development in Land Administration – a TEMPUS project.* The main aim of this project is the creation of training and education for the continuing professional development in Land Administration that are delivered as short cycle staff development activities. Since the programme will not be offered as part of an overall academic programme, it can be more flexible and can be targeted in different ways towards differing levels of ability and staff requirements.

Education amongst the rural population is also very important so that the people who are going to be actively involved and affected by the process of land consolidation will be in a position to understand the process and to make informed judgements. This concept will be invasive throughout the remainder of the document.

**Design considerations for land consolidation**

One of the most important considerations in any land consolidation process is the spatial data that will be used during the design and implementation of the scheme. Spatial data techniques, in particular orthophoto mosaics, digital thematic mapping, digital cadastres and topographic maps all provide the technical fields associated with resource planning and create a powerful means of discussing land use and land administration across decision-making boundaries at all levels (Riddell and Rembold, 2002). They have proved to be effective tools for community participation since they enable all the participants to define their spatial concepts and have these represented as an overlay on the specialist maps. It is their space that is being discussed and it is within their concept of this space that the land consolidation process with be implemented.

One of the most important tools of spatial data infrastructures is that of GIS/LIS, which can act as a means of cutting across conceptual frameworks and enable people to see the relationship between their land parcels, the land use planners agro-ecological zones, and the land administrator's environmental protection areas. It provides the basis for a dialogue between the concerns of the land user, the planner, and the administrator. For this to act in an effective manner, the implementation of an accurate and current land databank is essential.

The new participatory tools have also revolutionised a further critical aspect of land consolidation, the presentation of alternative solutions. The ability of presenting alternative solutions is part of the interactive bargaining process in which land users are able to view all possibilities, the effect of planning restrictions, policies and directives, to see the impact of their own proposals, change their proposals, and to consider the implications.

*A technical co-operation project in Hungary concerning land consolidation, village improvement and rural development*

The Hungarian Ministry of Agriculture, represented by the Departments of Land and Mapping and Rural Development requested FAO assistance to design, develop and test a strategy for land consolidation, village improvement and rural development in appropriate and representative locations in Hungary. This will be used to integrate experiences, lessons learned and best practices from previous interventions. This is the TAMA land consolidation and village improvement project, as discussed by Nyiri and Dixon-Gough (1999).

The main inputs from FAO include:

- preparing the conceptual and methodological guidelines for participatory land consolidation and village development programmes in Hungary;
- designing and developing an operative framework for rural development, land consolidation and village improvement projects, especially with relation to the forthcoming programmes of EC accession;
- assessing and validating the process of data collection, processing, transformation and application concerning tenure security, property rights, land consolidation, geographic and spatial information;
- consolidating the institutional, organisational and managerial framework for interdisciplinary cross-sector and multifunctional ‘One Stop Shopping’ scheme for local/regional rural institutions including public-private partnerships;
- testing best practice experiences and lessons learned in predominately agriculture based regions;
- designing and developing complementary research studies, and analysis to assess and appraise the potential economic, social and ecological impacts of land consolidation and village improvement in Hungary.

*Challenges to overcome unfavourable farming structure and to revitalise rural development*

The establishment of an efficient land administration system is the key for designing consolidation schemes for fragmented farming systems (Graefen, 2002). This would assist in the emerging of a medium sized commercial farming sector with economically viable individual and efficient corporate farms.

Many cadastral systems currently being developed on the basis of advice promote the multi-purpose cadastres instead of preparing a sound and accurate base for a simple cadastre that has been designed to incorporate additional digital data whilst it is being developed. It has to be emphasised that an efficient cadastre is still the core for any operational land administration and functional land consolidation system. This implies hard work and requires experience, dedication, and sustained financial support over a long period, rather than a 'quick-fix' programme. Graefen (2002) suggested that:

- land consolidation using methods developed for Western Europe provide a clear choice and have been chosen for many pilot regions. However, the consolidation process is expansive, very time consuming and difficult to arrange in order to meet equality and democratic standards, balanced exchange of land and interests, and efficiency criteria;
- simplified alternatives exist, but how far are they recognising equality and accuracy? How far are they going to be accepted?
- is the enabling of lively rental markets a better choice to follow? They are progressing and might contribute much faster to intended allocation results than enabling the conditions for functional land markets through land consolidation.

*The current situation in Bavaria*

There are currently approximately 140,000 peasant farmers in Bavaria occupying relatively small plots having an average size of 1.2 ha, characterised by scattered parcels, especially in those regions where parcels are divided through hereditary purposes (Attenberger, 2002). Land consolidation is therefore essential in order to consolidate the scattered properties and to facilitate economic cultivation.

An integral part of this design process is to ensure that the plots are accessible by planning new tracks and roads that are capable of carrying modern farm machinery and equipment. These roads must be constructed in such a way as to make them suitable to both the landscape and to the ecology.

A further integral part of land consolidation is the restoration of watercourses, and wet land, especially the water courses that have been straightened in the past and wetlands that have been drained. By restoring the natural qualities of the water bodies, such as water meadows and riparian vegetation, they may be seen as part of the flood prevention measures for downstream areas and settlements, but will also serve as valuable habits for flora and fauna.

Attenberger (2002) provides an interesting case study, which examines public infrastructures, e.g., the building of new motorways, land consolidation can serve to satisfy several objectives:

- to avoid the compulsory purchase of land from only those land owners who are going to be directly affected;

- to repair the infrastructure of agricultural areas; and
- to consolidate scattered property in the area through which the motorway will pass.

In one example of motorway construction, the State Authorities acquired land in the defined area as a land consolidation scheme. This land was then transferred into:

- land for the new motorway;
- the site of a new recreation area;
- an area for afforestation;
- land for other public facilities; with
- the balance being returned as consolidated holdings to the farmers.

This example demonstrates one of the most important elements of land consolidation, namely to buy land, and to subsequently transfer this land to sites where it is needed yet protecting the rights of the landowners and relocating land to them in alternate locations equivalent in value to the land of the previous plots. Furthermore, the former owners are compensated for any loss of land that will be spread equally amongst land users through the area of the land consolidation scheme.

In this particular example, the area of the land consolidation scheme occupied a total of 591 ha. Out of this total, the State purchased 63 ha of land to be used in the following way:

- 48 ha for the new motorway and associated infrastructure;
- 13 ha for the new recreational area;
- 1 ha for nature and landscape conservation; and
- 1 ha for afforestation.

In much the same way, land can be purchased for the restoration of watercourses, and wetlands, and also for public facilities. Between 1996 and 2000, the State of Bavaria purchased a total of 6,400 ha. This land has been used for a number of purposes, ranging from the sale of land to peasant farmers to increase their land holdings, to improve the natural environment, public transport facilities, and the restoration of watercourses and wetlands.

The acquired land can also be used to create new building land, particularly community or low cost housing in rural areas, and for buildings to house handicrafts and industry. Villages are also included in the land consolidation procedure to ensure the sustainable development of rural areas. This takes the form of a general development plan for the village that includes conservation measures, design considerations and recommendations for new buildings, but also the economic, ecological and cultural situation of the settlement.

Since structural changes in agriculture are characterised by the redundancy or displacement of traditional buildings (dwellings as well as farm buildings), land consolidation plans can be used to suggest new uses for those buildings, and also to restrict the development of unsuitable buildings (the TROPHY house).

It is accepted that the most important part of land consolidation is the public participation of those involved in the scheme. They are generally willing to engage in the communal tasks and to contribute towards the scheme providing they have the ability to influence the plans and decisions concerning their living space. However, useful participation and co-operation is not always easy and there is a case for the 'education' of the public for those areas likely to be the subject of land consolidation schemes. This will ensure that village renewal is more than simply construction but will also concern the character of the village and village life.

### *Design considerations for landscape elements*

Many of the landscape elements are modified over periods of time and the function they fulfilled in former agricultural systems has been superseded to accommodate the continued modernisation of agriculture (Kristensen et al., 2001). Examples can be cited from across Europe, from the 'prairie landscapes' of eastern England to the collective farms of Central and Eastern Europe.

Within the EU, there have been two driving trends:

- the requirement for greater efficiency by the farming sector. This has been referred to as the 'productivist phase'. To some extent, this is being superseded by the 'post-productivist' phase (Bowler and Ilbery, 1999). This phase is characterised by the emergence of new forms of agricultural systems that have more diversified objectives than those relating to intensive production methods;
- through the 'greening' of EU legislation, a greater awareness of the environmental impact of farming, and an increased number of part-time or hobby farmers who are able to pursue particular farming techniques. Much of this 'greening' process can be applied through land consolidation (as in the case of the Netherlands) or through 'urban farms' located within the urban fringe (as is frequently practised in Poland).

These elements virtually relate to a restructuring of the agricultural sector, which may be expressed in the form of a diversification of farming systems in different areas, according to the biophysical and socio-economic conditions (Sørensen and Pinto-Correia, 1998). Ilbery and Bowler (1998) have stressed that the post-productivist approach to farming does not mean the end of intensive farming, but rather to the co-existence of this type of production with more extensive types of farming: both of which will be complementary. It is naturally to be expected that there will be some degree of variation between farm systems and landscape activities in different areas, according to the local potential and constraints as a result of the different underlying landscape elements. It is possible to establish that specific relationships between landscape characteristics and farming activities can be identified in different administrative or sub-administrative areas (Halfacree and Boyle, 1999). Programmes of land consolidation should, and must, address these issues.

## Functional design and design support

In general, the term land consolidation is misleading since it suggests an exclusive orientation towards the area of land administration. Access to land and secure property rights are crucial but equally important are rural development elements such as local capacity building, rural services, infrastructure, employment schemes etc., in order to improve rural livelihoods (Riddell and Rembold, 2002). For example:

- in many transition countries in the CEE region, land reform in terms of restitution, compensation and distribution led to extreme land fragmentation, which hampers sustainable rural development and affects all sectors with agriculture obviously hardest hit;
- despite the fact that land fragmentation is widely known and recognized there are few interventions so far and no comprehensive study/research in CEE/CIS has been carried out;
- since the impacts and implications of land consolidation programmes are still very much a black box careful appraisal is needed to avoid/mitigate/forecast negative externalities for rural society;
- land consolidation represents a essential requisite for further interventions in the rural space especially for EU accession countries;
- complex land consolidation processes provide an excellent opportunity to integrate land tenure services into the broader framework of rural, regional development and substantial synergy effects for instance joint activities concerning rural institutions can be expected;
- lessons learned in Western Europe show clearly that for land consolidation participation is absolutely necessary and its implementation proved successful only where stakeholders and beneficiaries are involved in decision-making processes and existing, informal approaches and schemes are recognized and integrated into local democratic governance institutions.

To summarise these arguments, land fragmentation is one of the major obstacles to achieving sustainable rural livelihoods, especially in transition countries.

### *An example from Eastern Germany*

The need for land consolidation in most transitional countries is very high (Drees, 2002). One example from the former East Germany is the district of Neuruppin/Brandenburg that has an area of approximately 643,000 ha. Of this, some 60 per cent is agriculturally productive (398,000 ha) and it has been estimated by the Department of Land Consolidation that measures of land consolidation are needed for approximately 80 per cent of agricultural land. To address the increasing problems of rural poverty and deprivation, this should be addressed as soon as possible and as cheaply as possible. Since reunification, 13 land consolidation schemes have been prepared or are in the process of preparation, covering an area of 34,000 ha, representing about 10 per cent of the actual need.

An example of one project is that of the land consolidation programme for the commune of Meyenburg, an area of 4,875 ha, with 850 participants. The planning for this scheme started in 1993 and was finally completed in 2001. The scheme included a range of projects including the creation of an up-to-date cadastre; re-allocation of land; and development of a land use plan. The reorganisation of the land also provided the opportunity to design and implement ecological and infrastructural measures.

In the Meyenburg programme, many small property owners gave up their rights to their land in return for financial compensation. This made sections of land available to a 'land bank' to be used for land exchange, ecological, infrastructural improvements, and also to increase the land area of existing holders. Within the Meyenburg project one particular nature conservation project was undertaken to improve the riparian zone. Before the scheme was implemented, the parcels stretched to the bank of the river. By utilising the land available in the 'land bank', an area along the banks of the river have been handed over to municipality for the protection of the riparian environment of this stretch of water. In a similar manner, the featureless agricultural landscape has been improved through the creation of landscape shaping elements, such as trees and bushes.

In the case of the Meyenburg project, the total cost of the management process was about 13 per cent of the total land value. To this must be added the costs of the survey (approximately 500 euros per ha), which came to about 20 per cent of the value of the land. The total costs of 33 per cent of the value of the land were similar to other projects and ways must be found to reduce the overall costs.

One of the largest contributing factors to the high surveying costs was the high level of accuracy and precision that was originally specified. However, the surveying standards could be reduced without any real loss of accuracy since the boundaries are fixed to a precision of a 2 cms. If those boundaries were to be fixed by photogrammetric methods, the fieldwork can be reduced with a precision of about 7 cms.

This approach has been adopted in a project in the commune of Lenzen, in which a land consolidation programme was implemented over an area of some 3,600 ha. One of the objectives of this programme was to shift a dyke along the river Elbe to re-open natural flood areas. The surveying was carried out using photogrammetry and the total survey costs were reduced to 10 per cent of the land value.

## Conclusion

Land consolidation is of high political importance in most of the transitional countries and is under pressure of time. To fulfil these demands, land consolidation must be achieved in the shortest possible time, and as cheaply as possible. To achieve this, traditional standards must be lowered.

One of the greatest questions of land consolidation in a rural context is whether it is both affordable and feasible. Given the significant demographic changes in rural areas and the net migration from rural to urban areas, it can be argued that land consolidation as a result of fragmentation will eventually diminish as the value of the rural land declines. As in the case of the jurisdictions of the British Isles, where land consolidation as not been practised since the end of the last Enclosure movement, it is likely that market forces will be the principal driver of land consolidation in most parts of Europe.

The situation concerning urban areas is far more acute, given changes in industrialisation and the need for extra housing to satisfy the demands of those who have moved from rural areas. Here, the driver will be urban land consolidation and the need to regenerate former industrial sites and, at the same time, expand into adjacent peri-urban areas.

It is likely, therefore, that over the next decade the emphasis in Europe will turn from rural land consolidation to urban land readjustment.

## References

Attenberger, J., 2002. The right mix of instruments – land consolidation, land management, and land banking in Bavaria, Paper presented at the International Symposium on *Land Fragmentation and Land Consolidation in CEEC: A Gate Towards Sustainable Rural Development in the New Millennium*, Munich.

Bowler, I.R. and Ilbery, B.W., 1999. Agricultural land use and landscape change under the post-productivist transition: examples from the UK. In: Krönert, R., Baudrey, J., Bowler, I.R. and Reenberg, A. (eds.), *Land Use Changes and Their Environmental Impact in Rural Areas in Europe*, 121-140, Unesco, Paris.

Dixon-Gough, R.W., Heine, E. and Mansberger, R., 2002. Cadastre – Aspects and Procedures in Austria and the United Kingdom, *1st Congress on Cadastre in the European Union 15th-17th May, 2002*, Granada.

Drees. A., 2002. Challenges in East German states, Paper presented at the International Symposium on *Land Fragmentation and Land Consolidation in CEEC: A Gate Towards Sustainable Rural Development in the New Millennium*, Munich.

FAO, 2003. *The Design of Land Consolidation Pilot Projects in Central and Eastern Europe*, Land Tenure Studies 6, Food and Agriculture Organization of the United Nations, Rome, 2003.

Fearnside, P.M., 2001. Land-tenure issues as factors in environmental destruction in Brazilian Amazon: the case of southern Pará, *World Development*, **29**(8), 1361-1372.

Graefen, C., 2002. Land reform and land fragmentation for rural development in the CEE/CIS countries, Paper presented at the International Symposium on *Land Fragmentation and Land Consolidation in CEEC: A Gate Towards Sustainable Rural Development in the New Millennium*, Munich.

Halfacree, K. and Boyle, P., 1999. Migration, rurality and the post-productivist countryside. In: Boyle, P. and Halfacre, K. (eds.), *Migration into Rural Areas: Theories and Issues*, 1-20, John Wiley, Chichester.

Hawerk, W., 1995. Grundbuch and cadastral systems in Germany, Austria and Switzerland, *Modern Cadastres and Cadastral Innovations*, Commission 7, FIG, Delft.

Ilbery, B.W. and Bowler, I.R., 1998. From agriculture productivism to post-productivism. In: Ilbery, B.W. (ed.), *The Geography of Rural Change*, 57-84, Longman, Harlow.

Kristensen, S.P., Thenail, C. and Kristensen, L., 2001. Farmers' involvement in landscape activities: an analysis of the relationship between farm location, farm characteristics and landscape changes in two study areas in Jutland, Denmark, *Journal of Environmental Management*, **61**, 301-318.

Mansberger, R. and Muggenhuber, G., 1999. Spatial data bases as tools for land use and development, *Presented paper at the FIG Commission 3 Annual Meeting and Seminar*, Budapest, Hungary.

Márkus, B., 1998. Building a knowledge pool for land administration, *UN ECE MOLA Meeting on Land Markets*, Budapest.

Márkus, B., 2001. Staff development in Hungarian land administration, *Computers, Environment and Urban Systems*, **25**, 389-405.

Nyiri, J. and Dixon-Gough, R.W., 1999. Conceptions of rural planning following land privatisation in Hungary. In: Dixon-Gough, R.W. (ed.), *Land Reform and Sustainable Development*, 138-159, Ashgate Publishing Ltd, Aldershot.

Riddell, J. and Rembold, F., 2000. Farm land rationalisation and land consolidation: strategies for multifunctional use of rural space in Eastern and Central Europe, *Quo Vadis International Conference*, FIG Working Week 2000, Prague.

Riddell, J. and Rembold, F., 2002. Farm land rationalisation and land consolidation: strategies for multifunctional use of rural space in Eastern and Central Europe, Paper presented at the International Symposium on *Land Fragmentation and Land Consolidation in CEEC: A Gate Towards Sustainable Rural Development in the New Millennium*, Munich.

Rognes, J. and Sky, P.K., 1998. *Mediation in the Norwegian Land Consolidation Courts*, Working Paper 14, North America Series, Land Tenure Centre, University of Wisconsin-Madison.

Sørensen, E.M. and Pinto-Correia, T., 1998. Analysing current changes in farm structures in two Danish parishes: types of farmers and their strategies, *Forest and Landscape Research*, **1**, 491-503.

Van Dijk, T, 2003a. *Dealing with Central European Land Fragmentation*, Uitgeverij Eburon, Delft.

Van Dijk, T., 2003b. Scenarios of Central European land fragmentation, *Land Use Policy*, **20**(2), 149-158.

Van Dijk, T., 2004. Land fragmentation in Central Europe: how and whether to use western experience. In: Deaking, M., Dixon-Gough, R.W. and Mansberger, R. (eds.), *Methodologies, Models and Instruments for Rural and Urban Land Management*, 35-48, Ashgate Publishing Ltd, Aldershot.

Wilkowski, W. and Sobolewska, K., 1998. Land consolidation in Poland with respect to elements of ecological and physical planning, *Commission 7, FIG Congress*, Brighton.

Williamson, I. and Ting, L., 2001. Land administration and cadastral trends – a framework for re-engineering, *Computers, Environment and Urban Systems*, **25**, 339-366.

# Chapter 12

# Land Tenure Reform in Scotland

Anthony Andrew

## Introduction

The Scottish legal system is distinct within the United Kingdom. It has its roots in Roman law, it draws on learned writers in Scots Law, it uses case law but is less bound by precedent and once again has a Parliament within Scotland to refresh it with statutory reform. Between the Union settlement of 1707 and the Scotland Act 1998 it had to rely on finding Parliamentary time at Westminster for reform. Since the devolution settlement, the Scottish Parliament has the legislative time to undertake the worthwhile but time consuming task of implementing an extensive programme of property law reform. Much groundwork had been undertaken by the Scottish Law Commission in a series of discussion papers and detailed reports on title conditions, real burdens, the conversion of long leases and the law of the tenement over the period 1990 to 2000. In parallel, the Land Reform Policy Group had been at work and produced its recommendations in 1999 for land reform touching on agriculture and crofting. Drawing upon this work and often using the Scottish Law Commission's draft bills, the Executive was able to put detailed legislative proposals before the Scottish Parliament quickly. The project was ambitious in scope and had a very high profile within Scotland.

The Executive's programme of property law reforms came into force on 28 November 2004. It has produced a modern and simplified framework for the ownership of property. The Abolition of Feudal Tenure etc. (Scotland) Act 2000 had great symbolic significance. It paved the way for a new method of holding land and the rules for that were set out in the Title Conditions (Scotland) Act 2003. The final stage in the programme of property law reform was the Tenements (Scotland) Act 2004 which clarified the law on tenements and provides a structure for the maintenance and management of tenements if this is not provided for in the title deeds. Legislation had already dealt with the anachronistic phenomenon known as leasehold casualties through the Leasehold Casualties (Scotland) Act 2001. After that, there followed the Scottish Law Commission Discussion Paper on the Conversion of Long leases (Scottish Law Commission, 2001) for consideration, although there is no commitment to introduce legislation on this subject. Following the work of the Housing Improvement Task Force set up in December 2000, which published its first report in 2002 (Scottish Executive, 2002a); there is a housing bill before the Scottish Parliament on the repair of housing in 2005. Together with property law reform and rural land reform, this work comprises a third strand of reform.

The Scottish Executive's Land Reform (Scotland) Act 2003 addresses particular problems of the Scottish countryside. Scotland has about one third of the UK landmass, one tenth of its population, and earns more of its GDP from farming and agriculture, so rural questions relating to fragile areas with sparse populations are more important than in England. The Act opens up responsible public access to rural land, opportunities for rural communities to purchase land when it comes up for sale, a crofting community right to buy a landlord's interest in land at any time and reform of Agricultural Holdings legislation. Within Scotland, these were high profile and, at times, controversial proposals addressing problems, which have concerned Scotland for years, but which have never found legislative time at Westminster.

## Abolition of the Feudal Tenure (Scotland) Act 2000 and Title Conditions (Scotland) Act 2003

On 9 June 2000, Royal Assent was received for the Abolition of Feudal Tenure etc. (Scotland) Act 2000 sounding the death knell for a system which has endured for about eight hundred years. The Act largely implemented the recommendations of the Scottish Law Commission Report on Abolition of the Feudal System (Scot Law Com No 168).

The Act abolished the feudal system of land tenure and incorporated land previously held under a feudal superior into a system of simple, outright ownership. Feudal superiors lost their rights to collect feu duties and enforce most feudal real burdens. In other words, conditions and restrictions on the use of land imposed in feudal deeds have ceased to be enforceable by feudal superiors. Non residential lease lengths have been restricted to 175 years to prevent the Scottish leasehold system being adapted to replace the feudal system. Residential leases are already restricted to a term of no more than 20 years by the Land Tenure Reform (Scotland) Act 1974 section 8.

The feudal system developed in the twelfth century where the King, the paramount superior parcelled land out to powerful barons in exchange for military service. These in turn feued out the land to large landowners so that a chain of obligations was developed. At the end of the chain and at the base of the pyramid was the vassal who occupied the land in terms of the restrictions placed by the feu. By the nineteenth century, the obligations were mainly commuted to money. A landowner might develop a residential estate imposing a duty to build on the purchaser of each feu in conformity with an overall pattern and impose restrictions to preserve the amenity of the estate.

In addition, a feu duty could offer a landlord a stream of income. Over the last thirty years feu duties gradually disappeared. Inflation had reduced their real worth and the Land Tenure Reform (Scotland) Act 1974 provided for their compulsory redemption when a property was sold or for voluntary redemption at any time. However the need to obtain the feu superior's permission to effect minor changes such as sub-division into flats or an extension with a dormer window proved irksome to vassals particularly as the superior often charged money for the permission. The time had come to end this practice. However there was much that was valuable about the set up and so the Title Conditions Act preserves certain useful features.

During the industrial revolution, as towns and cities grew, plots were sold for housing with conditions in the title deeds specifying the type of building that could be erected and the subsequent use to which they could be put. They were effectively the private sector forerunner of the planning system. When the title deeds are registered in the Register of Sasines or the Land Register, the conditions become enforceable and are passed to each owner in succession. They can impose a positive duty to repair or restrict the use of a property to that of a single dwelling.

Flowing from its work on the feudal system, the Scottish Law Commission realised that the Law regulating restrictions on future land use embedded in the title deeds needed updating.

The most common restrictions or conditions are known as real burdens and they pass with the land. In October 1998, the Scottish Law Commission published a discussion paper (SLC, Discussion Paper No 106), and issued a Final Report on Real Burdens (Scot Law Com No 181) together with a draft Title Conditions (Scotland) Bill. The Title Conditions (Scotland) Act 2003 largely implemented the recommendations of the Scottish Law Commission.

The SLC looked critically at the role of real burdens to see if they still served a useful function given the post war planning system. They concluded that real burdens could regulate matters of details, which would simply swamp the public planning system. Many burdens continue to perform a valid function. Simple abolition would violate too many legitimate interests.

The Title Conditions (Scotland) Act 2003 complements and completes the process of feudal abolition. The Act provides a re-statement and clarification of the law on real burdens. It stipulates rules for the creation, enforcement, and extinction of real burdens. Burdens validly created under the old law remain valid burdens, though the Act provides a mechanism for getting rid of obsolete burdens. It should also be easier to find out who has the right to enforce burdens. The Act creates a framework for the way in which individuals may impose their own controls on property and default rules for a number of areas where property may not be fully regulated by title deeds. This is intended to improve the management of property to allow repair work to be carried out when required.

### *What burdens were preserved and what was provided for in the new legislation*

The most important types of burden that were saved are facility and service burdens. These burdens provide for essential matters such as maintenance and use of common facilities. The right of the superior to enforce has been transferred to those who now own the land, which was originally designed to benefit from the burdens.

It was recognised that some former superiors have a genuine interest as a neighbour in certain burdens, which protect their own land and preserve the character of their property. The legislation allowed the superior to save these burdens if the burdened land was within 100 metres of a building owned by the superior and used as a place for human habitation or resort. In order to save the burden the superior had to register it at the Registers of Scotland before 28 November 2004. Certain other rights such as pre-emptions and feudal sporting rights could also be preserved by registration.

Provision was also made for a new category of burden called personal real burdens. These, unlike other real burdens, are not created to benefit other land. The purpose is to protect valuable rights, for the benefit of the public. The 2000 Act allowed for the preservation of burdens that would otherwise have been lost when the feudal system was abolished, and the Title Conditions Act allowed them to be created in future. The names of these types of burden are: conservation burdens, maritime burdens, economic development burdens, rural housing burdens, and health care burdens. A conservation burden is one, which preserves the architectural or historical or other special characteristics of the land. They may only be enforced by conservation bodies, which must be designated by Scottish Ministers. Maritime burdens protect the public interest in the seabed and foreshore, and may only be enforced by the Crown. Economic development burdens may only be enforced by local authorities or Scottish Ministers, and may be created for the purpose of promoting economic development. Rural housing burdens are intended to provide rural housing and may only be enforced by rural housing bodies, which are designated in a list made by Scottish Ministers. Health care burdens may be created where land is being sold but it is intended that it should only be used for health care purposes. National Health Service trusts or Scottish Ministers may only enforce them. There are also manager burdens for the management of tenement type properties, which are mentioned later.

## Abolition of leasehold casualties

The Scottish Law Commission Report defines leasehold casualties as:

> A leasehold casualty is an extra amount, over and above the rent, which must be paid by the tenant to the landlord from time to time if the lease so requires. A true casualty is payable to the occurrence of an event. The word 'casualty' comes from the Latin *casus*, meaning in this context an event or occurrence. (Scottish Law Commission, 1998)

These created problems for residential tenants in some localities of Scotland where they appeared in long leases, in the same way feudal casualties created problems for vassals under the feudal system before they were abolished by the Feudal Casualties (Scotland) Act 1914. Landlords could demand a double rental payment or duplicate every time the lease changed hands on inheritance or in some cases every 19 years. Sometimes the sums involved were quite large, usually they were unexpected and involved legal fees. When individuals began to buy up estates with leasehold interests deliberately to exploit these archaic and forgotten provisions, public concern increased. These 'raiders of the lost titles' became a political issue and the then Secretary of State Donald Dewar in 1997 asked the Scottish Law Commission to examine the law. The resulting Commission report attached a draft bill to extinguish leasehold casualties and provide appropriate compensation. Eventually, after deliberation by the Scottish Parliament, the Leasehold Casualties (Scotland) Act 2001 received the Royal Assent on 12 April 2001.

## Tenement law reform

The Scottish Law Commission produced a discussion paper on the Law of the Tenement in 1990 (SLC Discussion Paper No 91) and the process bore fruit in their 1998 report on the Law of the Tenement (Sco Law Com No 162). The report and draft bill were complex and ran to some 200 pages, but the main recommendation was to remedy the lack of a robust management scheme. The Tenement (Scotland) Act 2004 largely implemented the recommendations of the Scottish Law Commission.

Tenements form over a quarter of the housing stock in Scotland and come in all shapes and sizes. Most tenements are residential blocks, but office blocks can also fall within the definition. This also applies to large houses that have been divided into flats.

Common law rules governing the maintenance and management of tenements have developed since the seventeenth century, but these were not comprehensive or without anomaly. The development of the law on real burdens, however, helped to impose obligations on successive owners to adhere to a detailed regime for management and repair of a tenement. These burdens were drawn up to suit the particular circumstances of the tenement. If title deeds made no provision on one matter, the common law applied on that one matter. The common law applied where there was a gap in the title deeds.

The Tenement (Scotland) Act 2004 forms the third and final part of the Executive's current programme of property law reform. The Act restates and clarifies the common law rules, which relate to the ownership of the various parts of a tenement. It also introduces a statutory management scheme called the Tenement Management Scheme, which provides a structure for owners to manage and maintain tenement property in circumstances where title deeds are silent or unworkable. The scheme includes provision for majority decision making on a range of issues (including having repairs carried out) if this is not provided for in the title deeds. The objective is to facilitate decision making so that flat owners can reach agreement about the best way of keeping their tenement in good condition. One recalcitrant owner will not be able to hold out against a majority decision to carry out repairs because a decision of the majority will be binding on all of the owners.

### *Long leases*

There have been concerns about certain types of long leases being used in a way that is analogous to a feudal tenurial relationship. In its 'Discussion Paper on conversion of Long Leases' the Scottish Law Commission says on page 1:

> Although important work remains to be done in other areas of property law, the present discussion paper marks the final stage of this structural review. In essence, it seeks to apply to certain long leases the principle of conversion already applied to feus by the Abolition of Feudal Tenure etc. (Scotland) Act. Under that, all feus are to be converted into ownership, on a day yet to be announced and known as the 'appointed day'. Feudal vassals will then become outright owners. The present paper seeks to extend that scheme to tenants holding under certain categories of long lease.

There is currently no commitment to legislate on this subject.

## Rural land reform

Following extensive deliberation, the Land Reform Policy Group published, in July 1999, a White Paper that contained innovative, yet controversial ideas to allow communities to register an interest in land, and to buy it when it comes up for sale. There followed unprecedented consultation. In February 2001, the Scottish Executive produced the Draft Land Reform (Scotland) Bill as a consultation paper. In addition, the Scottish Executive produced a Draft Scottish Outdoor Access Code for consultation. The latter proposed to clarify and widen public access rights to the countryside. The Act gave crofting communities rights to buy the landlord's interest in their crofting estate including certain valuable fishing rights, even where the landlord may be unwilling to do so.

The aim was to empower communities that wish to unite landlord and tenants' interests to do so with appropriate recompense to the landlord. This offers local rural communities greater control over the assets they have worked for years by giving them the initiative that proprietary land ownership gives. Already there have been community buy-outs on the Island of Eigg, in Assynt and on the Island of Harris, which is creating much interest.

In March 2005, a draft Crofting Reform (Scotland) Bill was laid before the Scottish Parliament to offer crofting communities greater rights and wider ability to determine their own future.

## Reform of agricultural leases

Since the immediate post war period, the relationship between agricultural landlord and tenant has been governed by the Agricultural Holdings (Scotland) Act 1948. This conferred security of tenure on the tenant except in fairly restricted circumstances usually involving bad husbandry or where the landlord had a development opportunity. The result was that landlords became reluctant to let new holdings and the percentage of tenanted land in Scotland dropped. As a vehicle to get around the provisions of the Act, partnership agreements developed to avoid the security of tenure problems. However, these were not seen as a universal answer.

The Land Reform Policy Group considered the suitability of existing agricultural holdings legislation to deal with modern needs as part of its work. There was also a desire to encourage more owners to let land. This led to the Scottish Executive producing a White Paper consultation document, 'Agricultural Holdings: Proposals for Legislation' in May 2000 followed by a draft bill in 2002, which received Royal Assent on 22 April 2003 and became the Agricultural Holdings (Scotland) Act 2003. This Act amends but does not replace the 1991 Act.

The key features of the 2003 Act are that it introduced new types of fixed duration tenancies. The Limited Duration Tenancy (LDT) has a minimum duration of 15 years, and the Short Limited Duration Tenancy (SLDT) with a maximum length of five years offer alternatives to the 1991 Act secure tenancy arrangements which are of indeterminate duration. The 2003 Act also provided new opportunities for tenants with a long-term interest in their holding to diversify into non-agricultural activities such as developing a bed and breakfast business or undertake agri-environment schemes. It also introduced less costly mechanisms for resolving disputes between landlord and tenant than the traditional procedures.

There is also a pre-emptive right to buy for secure agricultural tenants with 1991 Act secure tenancies to buy their holdings when the landlord decides to sell his interest.

England had similar post-war legislation and addressed the same problem through the Agricultural Tenancies Act 1995. Of this English legislation, which introduced farm business tenancies, it was said:

> New lettings will be governed by the 1995 Act. A precedent has been included for guidance. However, it should be stressed that the essence of the legislation is freedom of contract. Agreements should be drafted to meet the individual needs of landlords and tenants. (Sydenham, 1997)

This subject shows an interesting contrast in approaches between Scotland and England and Wales, which the devolved system of two Parliaments makes easier.

## Conclusion

The Scottish Parliament and the Scottish Executive have used the extra legislative time available with the new Scottish Parliament creatively to completely update the system of Scottish land tenure, and develop solutions specifically tailored to Scottish needs in a way that would have been impossible with a single legislature.

## References and Acts referred to in the text

Abolition of Feudal Tenure (Scotland) etc Act 2000 asp 5.
Agricultural Tenancies Act 1995 c 8.
Agricultural Holdings (Scotland) Act 2003 asp 11.
Feudal Casualties (Scotland) Act 1914 c48.
Land Reform (Scotland) Act 2003 asp 2.
Land Tenure Reform (Scotland) Act 1974 c.38
Leasehold Casualties (Scotland) Act 2001 asp 5.
Scotland Act 1998 c 46.
Scottish Executive, 2000. *Agricultural Holdings Proposals for Legislation*, May 2000, Scottish Executive, Edinburgh.
Scottish Executive, 2001. *Land Reform the Draft Bill*, Consultation Paper, February 2001, Scottish Executive, Edinburgh.
Scottish Executive, 2001. *A Draft Scottish Outdoor Access Code*, February 2001, Scottish Executive, Edinburgh.
Scottish Executive, 2001. *Title Conditions (Scotland) Bill*, Consultation Paper, May 2001, Scottish Executive, Edinburgh.
Scottish Executive, 2001. *Title Conditions (Scotland) Bill*, Scottish Law Commission Draft, Edinburgh.
Scottish Executive, 2002a. *Issues in improving Quality in Private Houses: the First Report of the housing Improvement Task Force*, Consultation Paper, Scottish Executive, Edinburgh.
Scottish Executive, 2002. *Crofting Reform Proposals for Legislation*, Scottish Executive, Edinburgh.
Scottish Executive, 2005. *Draft Crofting Reform (Scotland) Bill*, Scottish Executive, Edinburgh.
Scottish Law Commission, 1990. *Law of the Tenement*, Scottish Law Commission Discussion Paper No. 91, Edinburgh.
Scottish Law Commission, 1998. *Discussion Paper on Real Burdens*, Scottish Law Commission Discussion Paper No. 106, Edinburgh.
Scottish Law Commission, 1998. *Report on the Law of the Tenement*, Scottish Law Commission Discussion Paper No. 162, Edinburgh.
Scottish Law Commission, 1998. *Report on Leasehold Casualties*, Scottish Law Commission Discussion Paper No 165, Edinburgh.
Scottish Law Commission, 2000. *Report on Real Burdens*, Scottish Law Commission Discussion Paper No 181, Edinburgh.
Scottish Law Commission, 2001. *Discussion Paper on Conversion of Long Leases*, Scottish Law Commission Discussion Paper No. 112, Edinburgh.
Scottish Law Commission, 1998. *Discussion Paper on Real Burdens*, Scottish Law Commission Discussion Paper, Edinburgh.
Sydenham, A., 1997. *Agricultural Tenancies*, Cavendish Publishing Ltd., London.
Title Conditions (Scotland) Act 2003 asp 9.

The kind assistance of the Scottish Executive Justice Department and the Scottish Executive Environment and Rural Affairs Department is gratefully acknowledged in the preparation of this chapter. Responsibility for all the views and any errors remain the responsibility of the author.

# Index